ISBN 978-0-266-88094-3
PIBN 10903806

ESSENTIALS

OF

MATERIA MEDICA

PHARMACY

AND

PRESCRIPTION WRITING

ARRANGED IN CONFORMITY WITH THE CLASSIFICATION IN THE LAST
EDITION OF PROF. H. C. WOOD'S "THERAPEUTICS" AND
FOLLOWING THE COURSE OF PHARMACY AS TAUGHT
IN THE UNIVERSITY OF PENNSYLVANIA

BY

EDWIN A. HELLER, M.D.

QUIZ-MASTER IN MATERIA MEDICA AND PHARMACY AT THE MEDICAL
INSTITUTE, UNIVERSITY OF PENNSYLVANIA

PHILADELPHIA

P. BLAKISTON, SON & CO.

1012 WALNUT STREET

1897

By the kind permission of Prof. H. C. Wood the author has been enabled to prepare the following volume. The definitions are mostly those given by the last edition of the United States Pharmacopœia. That part of the work relating to pharmacy has been arranged with special reference to the course as given at the University of Pennsylvania. Special attention has been given to the metric system, heretofore too much neglected.

The author trusts that this little work will lighten the labors co-existent with the entrance into the study of materia medica, and if it accomplishes this end he will feel amply repaid for the care and exertion required to compile it.

EDWIN A. HELLER.

934 FRANKLIN STREET, PHILADELPHIA,
July, 1897.

iii

CONTENTS.

CHAPTER VI.

CHAPTER VII.

CHAPTER VIII.

PART II.

CHAPTER I.

CHAPTER II.

CHAPTER III.

CHAPTER IV.

ADDENDA.

viii

ERRATA.

Page 115, first line, should read, "Solutions of gun-cotton in ether.

Page 131, third line from bottom, "contain" should be "contains."

Page 167, tenth line from bottom, "or" should be "and."

PART I.

PRESCRIPTION WRITING.

CHAPTER I.

DEFINITIONS.—PARTS OF A PRESCRIPTION.

The word prescription (from the Latin *præ*, before, and *scriptum*, perfect participle of *scribo*, to write, meaning, written,), was at one time understood to mean any direction whatever, either verbal or written, given to the patient. It included directions as to diet, ventilation, heat, light of his apartment, etc.; in fact, any direction whatever relating to the care of the patient or his surroundings.

At present, however, a prescription is generally understood to be a written formula containing the names and quantities of a drug or drugs, together with directions to the apothecary for dispensing, and directing also the patient as to the manner, method, and frequency of administration.

A prescription may be either (1) simple or (2) compound; and the formula it contains may be

either (1) officinal or (2) extemporaneous or magistral.

A **simple prescription** contains but one ingredient.

Compound prescriptions always contain more than one ingredient.

An **officinal preparation** or formula is one published and authorized by the United States Pharmacopeia, and although it may contain numerous ingredients, in prescribing it is necessary to write only the officinal name, together with the dose and directions.

An **extemporaneous** or **magistral formula** is one composed by the physician to suit the individual case.

A typical prescription consists of:

1. The **superscription**, heading.

2. The **inscription**, names and quantities of the drugs prescribed.

3. The **subscription**, directions to the pharmacist.

4. The **signature**, directions to patient.

5. **Date, and the signature** of the physician.

I. In English the *superscription* is always the symbol ℞; a combination of R from *recipe* (Latin, imperative of *recipio*), take, and the zodiac sign ♃. Originally, prescriptions were always begun with an invocation to Jove or Jupiter, and his blessing invoked on the action of the remedy, whence we derive

the combination of the R and ♃. When Christianity supplanted the heathen beliefs, the prayers were abbreviated and changed in various ways: as, A, Ω, the beginning and end, the first and last, the alpha and omega of everything; JJ, Juvante Jesu (Jesus helping); JD, Juvante Deo (God helping); ND, Nomine Dei (in God's name); and at one time also the simple +, the sign of the cross. But all these have been discarded for the old ℞. In France P or Ps (*prenez*, take,) is employed.

II. The *inscription*, or body of the prescription, contains the names and quantities of the ingredients, and in a typical prescription includes:

1. The **basis**, the principal active agent.

2. The **adjuvant**, or auxiliary intended to aid and increase the action of the basis.

3. The **corrective** or **corrigent**, to correct or modify one or both of the above two.

4. The **vehicle** or **excipient**, to render palatable, assimilable, or easy of administration the entire prescription.

The old maxim of Asclepiades, "*Curare cite tute et jucunde*," might be applied as follows:

Curare	(cure),	with the	*Basis.*
Cite	(quickly),	" "	*Adjuvant.*
Tute	(safely),	" "	*Corrective.*
et Jucunde	(pleasantly),	" "	*Vehicle.*

The names of the ingredients are always written in

Latin and are in the *genitive case*, governed by *recipe*.

The quantities, if written out in Latin, which, however, is practically never done, are always in the *accusative*.

The *subscription* or instruction as to method of dispensing is always in Latin. (A list of the principal phrases will be found on pages 61–65.)

The *signature*, written Signa, or Sig., consists of the directions to the patient, is always in English, and should always be as simple and distinct as it possibly can be written. Even in cases where it is desired to conceal from the pharmacist the purpose for which the remedy is intended, this can be done at no sacrifice of meaning; *e. g.*, in case an injection should be ordered we may simply sign, " Use as wash ; " " Bathe affected part two or three times daily," etc. Thus the druggist is often unable to tell if the prescription is intended for eye, mouth, vagina, urethra, or other part of the body, and the patient protected both while procuring his prescription and at home, should the preparation be seen by those who have no right to do so.

CHAPTER II.

APOTHECARIES' WEIGHT AND MEASURE.

At present, although the metric system is really the only system founded on a definite scientific basis, both the apothecaries' weight and measure and the metric system are employed.

APOTHECARIES' WEIGHT.

20 grains (gr.) = 1 scruple.
60 grains or 3 scruples (Ɉ) = 1 dram.
480 grains or 8 drams (ʒ) = 1 ounce.
5760 grains or 96 drams or 12 ounces (℥) = 1 pound.

. In Latin, respectively:

Pound,	symbol	℔,	=	libra.
Ounce,	"	℥,	=	uncia.
Dram,	"	ʒ,	=	drachma.
Scruple,	"	Ɉ,		scrupulum.
Grain,	"	gr.,		granum.

The scruple (Ɉ) is practically obsolete, because if not carefully written it is easily confounded with the dram (ʒ), and thus may give rise to serious consequences. Amounts less than one dram should be expressed in *grains*.

The British pharmacopeia directs that the pound should contain 16 ounces, each of them equaling 437.5 grains, or 7000 grains to the pound, being

18   FOR THERAPEUTIC NOTES.

considerably more than our pound.   This must be borne in mind in the use of formulæ based on the British standard.

APOTHECARIES' OR WINE MEASURE.

60 minims (m̴)   = 1 fluidram.
8 fluidrams (f ʒ)   = 1 fluidounce, or 8 f ʒ, or 480 m̴.
16 fluidounces (f ℥) = 1 pint, or 16 f ℥, or 128 f ʒ, or 7680 m̴.
8 pints (O)   = 1 gallon (C), or 8 O, or 128 f ℥, or
                        1024 f ʒ, or 61,440 m̴.

In Latin, respectively:

The gallon,   symbol C, = congius.
" pint,   "   O, = octarius.
" fluidounce, "   f ℥, = fluiduncia
" fluidram,   "   f ʒ, = fluidrachma.
" minim,   "   m̴, = minimum.

The English *pint* contains 20 fluidounces, and the fluidounce equals 7 fluidrams and 2½ minims; their *minim*, therefore, is equal only to .96 of ours, theirs weighing approximately .91 of a grain, while ours is equivalent to .95 of a grain.

As practically all patients are unfamiliar with apothecaries' measure, we must employ some domestic measure fairly equivalent to it, and for that purpose we generally consider—

The drop   = a minim.
" teaspoon   = a fluidram.
" dessertspoon   = 2 fluidrams.
" tablespoon   = 4 fluidrams, or ½ of a fluidounce.
" wineglass   = 2 fluidounces.
" teacup   = 4 fluidounces.

Naturally, it is at once evident that as the drops of some liquids are much larger than those of others, and as teaspoons vary greatly in capacity, to say nothing of the balance, the method is faulty and not to be relied upon for accurate dosage.

In such cases it is necessary to procure an accurate graduate at once and use only this in giving the medicine to the patient.

If drops are ordered (as for use in eye work), we may order a pipet. Should we desire to give only minims in this manner, accurate minim pipets may be used. That the size of drops varies greatly may easily be seen by glancing at the following list of drugs. Drops in one fluidram:

Dilute sulphuric acid	=	54 to 48
Aromatic sulphuric acid	=	115 " 145
Hydrocyanic acid	=	45 " 48
Ether	=	150 " 160
Chloroform	=	180 " 275
Alcohol	=	120 " 143
Castor oil	=	50 " 58
Tinct. aconite	—	115 " 125
Tinct. opium	—	105 " 145

CHAPTER III.

METRIC SYSTEM.

The **metric system** now used practically exclusively in France and Germany is being rapidly adopted in the other countries of Europe, and is making rapid progress, as it deserves to, in this country. It is the only rational system of weights and measures we possess, the unit of length being the, **meter**, which equals one forty-millionth of the earth's circumference through the poles, or one ten-millionth of the distance from the pole to the equator. It equals 39.37 inches, being 3⅓ inches (about) more than our yard.

The **gram**, the unit of weight, is the weight of one cubic centimeter (c. c.) of water at its greatest density (4° C. or 39° F.).

In writing a prescription according to the metric system, if we desire all the ingredients to be weighed, we merely place the symbol gm. above the figures. If, however, the liquids are to be measured, we write gm. and c. c. (cubic centimeters).

The gram (solid) equals 15.432 grains. The gram (of water) measures 16.231 minims.

23

The subdivisions of the units are formed by prefixing to the unit the Latin—

Milli (from *mille*) $= \frac{1}{1000}$ of the unit.
Centi (from *centum*) $= \frac{1}{100}$ of the unit.
Deci (from *decem*) $= \frac{1}{10}$ of the unit.

Thus :

Two decigrams $= \frac{2}{10}$ of a gram.
Or two millimeters $= \frac{2}{1000}$ of a meter.
Or one centiliter $= \frac{1}{100}$ of a liter.

The multiples of the units are formed by prefixing to the unit the Greek—

Deca   (from Δεκα) $=$ 10 times the unit.
Hecto (from Ἑκατον) $=$ 100 times the unit.
Kilo   (from Κίλιος) $=$ 1000 times the unit.
Myria (from Μυριας) $=$ 10,000 times the unit.

Thus :

A decaliter $=$ 10 liters.
A hectometer $=$ 100 meters.
A kilogram $=$ 1000 grams, etc.

Thus the multiple and subdivisions would be of the—

Gram.		*Meter.*		*Liter.*
Milligram,	$\frac{1}{1000}$	millimeter,	$\frac{1}{1000}$	milliliter.
Centigram,	$\frac{1}{100}$	centimeter,	$\frac{1}{100}$	centiliter.
Decigram,	$\frac{1}{10}$	decimeter,	$\frac{1}{10}$	deciliter.
Gram,		meter,		liter.
Decagram,	10	decameters,	10	decaliters.
Hectogram,	100	hectometers,	100	hectoliters.
Kilogram,	1000	kilometers,	1000	kiloliters.

1 gram, . . 15.432 grains.   1 meter, . . 39.37 inches.
1 liter, . . 2.113 pints.

Instead, however, of writing out in full the prefixes of the units, we employ the decimal system entirely; thus:

2	or	2
.125	"	125
.025		025

The whole numbers always signify grams and cubic centimeters, according to the symbol at the top; thus: gm. and c. c.   If, however, there is no symbol, then all the ingredients—liquid and solid—are supposed to be weighed out in grams.   Thus the above amounts would equal 2 grams; $\frac{125}{1000}$ of a gram or 12½ centigrams; $\frac{25}{1000}$ of a gram or 2½ centigrams or 25 milligrams.

Although it is sufficient to merely indicate the decimals by a point, as in the first example, in writing prescriptions it is policy to always use the line, as shown in the second example, so as to leave no possible room for doubt, as a spot in the paper may be mistaken for the point, and this would, of course, multiply or divide the result, possibly ten, possibly a hundred, fold.

For all practical purposes, and for converting apothecaries' weight into the metric, and *vice versâ*, a gram may be considered equal to 15 or 16 grains, using that number which divides or multiplies most easily.   Except in cases of poisons, alkaloids, and very powerful drugs, where it is well to consider that 15⅜ grains equal 1 gram.

Likewise, with the liquid preparations, either 15 or 16 minims may be considered equivalent to one cubic centimeter.

The fluidram may be said to equal 4 c. c. The dram to equal 4 grams. The fluidounce may be considered as equivalent to 30 or 32 c. c. ; and the ounce, 30 or 32 grams. The liter is equal to 2.113 pints.

It will be seen that the table for liquids is computed for water at 4° C. ; and, consequently, to be strictly accurate, allowance would have to be made for the weight of all the liquid preparations compared to that of water. Thus, the dose of all liquids lighter than water: *e. g.*, alcohol, ether, the tinctures, etc., would be slightly less ; the dose of heavier liquids : *e. g.*, the syrups, glycerites, decoctions, etc., would be larger in order to be absolutely correct. But practically the difference is so small that in most, if not all, cases it may be discarded.

TABLE TO CONVERT METRIC INTO APOTHECARIES' WEIGHT, AND *Vice Versâ*.

$$
\begin{array}{rll}
\text{1 grain} = & .06 \text{ gram.} \\
\ni j = \text{20 grains} = & 1.2 \ \text{ grams.} \\
\chi j = 60 \quad `` & - 4. \quad `` \\
\zeta j = 480 \quad `` = 30. & \text{or 32 grams.} \\
\text{1} \ m = & .06 \text{ c. c.} \\
f \chi j = 60 \ m = 4. & `` \\
f \zeta j = 480 \ m = 30. & \text{or 32 c. c.}
\end{array}
$$

It will be seen that the figures do not accord ex-

actly with the results of multiplication, but we even them up in order to obviate the fractions.

1.	gram	$=$	15. or 16 grains.
.1	"	$=$	1.5 grains.
.01	"	$=$	.15 grain.

1.	c. c.	$=$	16. $\mathfrak{m}$.
.1	c. c.	$=$	1.6 $\mathfrak{m}$.
.01	c. c.	$=$	.16 $\mathfrak{m}$.

TABLE FOR CONVERTING APOTHECARIES' WEIGHT INTO THE
METRIC SYSTEM, AND *Vice Versâ*.

SOLIDS.		LIQUIDS.	
Apothecaries'.	Metric.	Apothecaries'.	Metric.
GRAINS.	GRAMS.	MINIMS.	GRMS AND C.C.
$\frac{1}{64}$	.001	ᵥ	.06
$\frac{1}{40}$	.0015		.12
$\frac{1}{30}$	.002	3	.18
$\frac{1}{20}$	.003	4	.24
$\frac{1}{16}$	.004	5	.3
$\frac{1}{12}$	.005	6	.36
$\frac{1}{10}$	.006	7	.42
$\frac{1}{8}$	.008	8	.5
$\frac{1}{4}$	.016	9	.55
$\frac{1}{3}$	.02	10	.6
$\frac{1}{2}$	.03	12	.72
1	.065	16	1.
2	.013	20	1.25
3	.2	25	1.55
4	.26	48	3.
5	.32	50	3.12
10	.65		
15	1.	(f℥) 60	3.75
(℈) 20	1.3	240	15.
30	1.95	(f℥) 480	30.
(℥) 60	3.75		
(℥) 480	30.		

CHAPTER IV.

METHOD OF WRITING PRESCRIPTIONS.— CONVERSION OF APOTHECARIES' WEIGHT AND MEASURE INTO THE METRIC SYSTEM, AND VICE VERSA.

The proper, as well as the quickest and safest, way to write a prescription is to put down the names of the ingredients intended to be used ; then determine the number of doses to be given *in toto;* and, finally, after multiplying the individual dose of each ingredient by the number of doses, put the corresponding amount opposite each drug.

Example : If we wish to write for the compound cathartic pill of the United States Pharmacopeia :

		Gm.
℞ .	Extracti colocynthidis compositi, ...	8.
	Hydrargyri chloridi mitis, ...	6.
	Extracti jalapæ, ...	3.
	Cambogiæ, ...	15.
	Aquæ, quantum sufficiat.	
M.	Ft. pilulæ No. x.	

Having decided to give 10 pills, and deciding the single dose of colocynth to be 8 centigrams, we multiply 8 centigrams by 10, equaling 8 decigrams, which we put opposite the colocynth.

The single dose of calomel we are going to give

being 6 centigrams, we multiply this by 10, equaling 6 decigrams, which we then place opposite its line, and so on.

The quantities of the drugs in metric prescriptions are expressed always in the Arabic numerals; while in the apothecaries' weight we use the Roman numerals, except in the case of fractions, where, for greater accuracy, we use ordinary figures; or in cases where a large or ordinarily poisonous dose is intended we may place the Arabic numeral in parenthesis alongside the Roman, in order to assure the druggist that a large amount is intended; thus f℥iij (3).

For converting apothecaries' weight into the metric, or *vice versâ*, reference to the table at the end of chapter III will obviate the necessity of multiplication and division. It is, however, advisable, for the sake of practice, that the student convert several prescriptions without the table, in order to be familiar with the method and able at any time to convert one table into the other. For example, to convert the following into the metric system:

For JOHN SMITH.

℞. Pulveris extracti glycyrrhizæ,
      "   acaciæ,   āā   gr. viij.
      "   sacchari,   gr. x.
      "   kino,   gr. ij.
   Spiritus aromatici,   ℥v.
   Mellis despumati, q. s.
M.   Ft. massa, in pilulas numero triginta dividenda.
SIGNA.—One after meals.

601 LANCET AVE., *Nov. 15, 1896.* JOHN MEDICUS, M.D.

Considering 1 gram to equal 16 grains (see chap. III) we find the first and second ingredients to equal ¼ of a gram, which we would write | 5. The third equals $\frac{10}{16} = \frac{5}{8}$ or | 625 grams or 62½ centigrams.

The fourth equals $\frac{2}{16} = \frac{1}{8}$ or | 125 grams or 12½ centigrams. The last quantity, 5 ♏, would be expressed in cubic centimeters, and as we may count either 15 or 16 minims as equaling 1 c. c., we will take 15 ; therefore, we would have $\frac{5}{15}$ or ⅓ of a cubic centimeter or 3⅓ cubic decometers, written | 33.

The body of the prescription then would read :

		gm.	c. c.
℞.	Pulveris extracti glycyrrhizæ,		
"	acaciæ, aa	5	
"	sacchari,	625	
"	kino,	125	
	Spiritus aromatici,	33	
	Mellis despumati, q. s. •		

Or, *vice versâ*, to convert the following into apothecaries' weight :

For SAM SMALL.

℞.	Potassii bromidi,	15	
	Antipyrin,	7	5
	Acidi arsenosi,	06	
M.	Fiant pulveres numero viginti.		

SIGNA.—One at bedtime.

1428 ENGELY ST., *Nov. 15, 1896*.   THOS. JONES, M.D.

Now, 1 gram equals 15 or 16 grains ; 15 (grams) × 16 (grains) = 240 ; 240 grains ÷ 60 (the number of grains in a dram) = 4 ; consequently the total is ℨiv.

The second ingredient calls for 7.5 grams: $7\frac{1}{2}$ (grams) $\times$ 16 (grains) $=$ 120; 120 (grains) $\div$ 60 (grains in a dram) $=$ 2, hence ℥ij. The third quantity: .06 grams or 6 centigrams $= \frac{6}{100}$ grams $\times$ 16 (the number of grains in a gram) $= \frac{96}{100}$; practically, 1 grain. Of course, there is a difference of $\frac{4}{100}$ grains, but as there are 20 doses, the difference in each dose would be but $\frac{4}{2000}$ of a grain, which is insignificant and may be disregarded. In fact, it is the rule in transposing from one system to another, to always even up the amounts: unless the drug is extremely potent, it will be found that the difference one way or the other will be too small to be of practical import. And the prescription then would read:

For SAM SMALL.

℞.   Potassii bromidi,   ℥ iv.

       Antipyrin,   ℥ ij.

       Acidi arsenosi,   gr. j.

M.   Fiant pulveres numero viginti.

SIGNA.—One at bedtime.

1428 EDGELY ST., *Nov. 15, 1896.*   THOS. JONES, M.D.

THE GRAMMATICAL CONSTRUCTION OF PRESCRIPTIONS.

Latin is the language *par excellence* for prescrip-
ns. Although the physician may use English or
y other language if he sees fit, the arguments in
or of Latin far outweigh any which may be
)ught against it.

First, it is a "dead" language, does not undergo
y change, and words expressed in Latin are under-
od all over the civilized world, whereas if we
ote prescriptions only in the current tongue, spe-
l knowledge of that language would be necessary
translate it into any other language. The com-
:hension, however, of even a very moderate
)ount of Latin enables us to understand a pre-
iption written in any of the civilized countries.
tin is the universal language of science,—the
tanic and chemic names of all our remedies are
Latin.

When we express the name of a drug in Latin it
ers distinctly and positively to only one drug,
ereas the English word may include a number
drugs entirely different from one another. Thus,

43

cimicifuga means only the cimicifuga racemosa, whereas the English name "snakeroot" is applied to numerous plants, each differing according to its habitat. If we write Indian hemp, either apocynum cannabinum or cannabis indica may be dispensed. Wintergreen may mean either chimaphila or gaultheria.

Checkerberry may mean either uva ursi or gaultheria.

In any or all such cases the drug dispensed may have directly opposite effect on the patient from that desired, and not alone may it be inert, but it may be positively harmful, if not disastrous.

Again, there is a certain prejudice against the use of certain drugs,—as mercury and iodid of potassium,—and should the prescription be in English the patient may refuse to take it.

Again, it is sometimes desirable that a patient should not know the exact nature of the drugs he is taking, or he may wish to take our prescription to a foreign country. In either case Latin fulfils all requirements.

The directions to the patient, however, should invariably be written in English. There is absolutely no valid reason why he should not read them; if they be written in Latin, should the patient forget the instructions—and patients are often singularly negligent in this respect—he would be at a loss how to take the remedy, and at what times.

certain rudimentary knowledge of Latin is
ssary in order to write a proper prescription.

he following few simple rules are indispensable:

. The names of the drugs are always in the
itive if the quantity is expressed (governed by
℞ (take) amount *of*).

. If no quantity is expressed, but a numeral ad-
ive follows, the noun is always in the accusative.

. The quantities are always in the accusative,
erned by the imperative *recipe* (take).

. Adjectives agree with their nouns in number,
der, and case.

'erbs, prepositions, conjunctions, and ad-
bs undergo no change, the principal ones
ig :

Verbs.	Abbreviations.	English.
:cipe,	℞,	take.
isce,	M.,	mix, or mix thou.
gna,	Sig.,	mark or label.
at,	Ft.,	let (it) be made.
ant,	Ft.,	let (them) be made.
ide,	Add.,	add.
illiat,	Bull.,	let boil.
la,	Col.,	strain.
ivide,	Div.,	divide.
acera,	Mac.,	macerate.
:petatur,	Rep.,	repeat.
lve,	Sol.,	dissolve.
fficit,	Suf.,	it suffices.
:re,	Ter.,	rub.

Prepositions, etc.

Cum, ...	with.
In, ...	in, into.
Ad, ...	to, up to.
Et, ...	and.
Ana, ... aa, ...	of each.

FIRST DECLENSION.

All pharmacopeial nouns ending in a belong to
the first declension, are feminine gender, and are
declined as follows, except aspidosperma, physostigma, and folia (plural):

Singular.		*Plural.*	
Nom., gutta (a drop),	–a.	Guttæ (drops),	–æ.
Gen., guttæ (of a drop),	–æ.	Guttarum (of drops),	–arum.
Acc., guttam (a drop),	–am.	Guttas (drops),	–as.

The stem (*gutt*) remains unchanged, and if the
case-endings are committed to memory there will be
no trouble in declining any noun.

(Aspidosperma and physostigma change to
aspidospermatis and physostigmatis in the
genitive. Folia (leaves) is plural; genitive, foliorum.)

Two pharmacopeial nouns of the first declension
end in e; *e. g.*, aloe and mastiche. The case-endings in the singular are: Nom., *e*; gen., *es*; acc., *en*.
The plural is like that of *gutta*.

SECOND DECLENSION.

All pharmacopeial nouns ending in us (except
Rhus spiritus, quercus, and fructus—fourth

declension) belong to the second declension.   They are nearly all masculine, declined thus :

Singular.		*Plural.*	
Nom., syrup*us* (a syrup),	–us.	Syrup*i* (syrups),	–i.
Gen., syrup*i* (of a syrup),	–i.	Syrup*orum* (of syrups),	–orum.
Acc., syrup*um* (a syrup),	–um.	Syrup*os* (syrups),	–os.

All pharmacopeial nouns ending in **um** are neuter, of second declension, and are declined thus :

Singular.		*Plural.*	
Nom., acet*um* (a vinegar),	–um.	Acet*a* (vinegars),	–a.
Gen., acet*i* (of a vinegar),	–i.	Acet*orum* (of vinegars), –orum.	
Acc., acet*um* (a vinegar),	–um.	Acet*a* (vinegars),	–a.

Rhus changes to **Rhois.**

THIRD DECLENSION.

All declinable pharmacopeial nouns whose nominative ends otherwise than in **a, us,** and **um,** are (except **aloes, mastiche, eriodictyon, hæmatoxylon,** and **toxicodendron**) of the third declension.   They are declined like **liquor. Digitalis,** however, does not change the genitive.   Nouns ending in **as** change to **atis ; as, acetas, acetatis ;** * **carbonas, carbonatis ; sulphas, sulphatis,** etc.

Singular.		*Plural.*	
Nom., liquor (a solution).		Liquor*es* (solution),	–es.
Gen., liquor*is* (of a solution), –is.		Liquor*um* (of solutions),	–um.
Acc., liquor*em* (a solution), –em.		Liquor*es* (of solutions),	–es.

---

* See note at end of chapter.

EXCEPTIONS.

Nom.,	anthemis,	*Gen.,*	anthemidis.
"	cortex,	"	corticis.
--	pepo,	"	peponis.
"	phosphis,	"	phosphitis.
"	sulphis,	"	sulphitis.
"	mucilago,	"	mucilaginis.
"	solidago,·	"	solidaginis.
"	colocynth,	"	colocynthidis.
"	hæmatoxylon,	"	hæmatoxyli.
-	radix,	"	radicis.
"	semen,	"	seminis.

FOURTH DECLENSION.

We employ only three nouns of the fourth declension; they are: **Fructus, spiritus** (masculine), and **quercus** (feminine), which are declined thus:

Singular.		*Plural.*	
Nom., spirit*us* (a spirit), –us.		Spirit*us* (spirits),	–us.
Gen., spirit*us* (of a spirit), –us.		Spirit*uum* (of spirits),	–uum.
Acc., spirit*um* (a spirit), –um.		Spirit*us* (spirits),	–us.

The ablative case is used only after **cum** (with); as :

Cum creta (with chalk).
" calce (with lime).
" ferro (with iron).
" semisse (with a half), etc.

The following nouns are indeclinable.   The genitive does not change :

Amyl.	Elixir.	Pyrogallol.
Azedarach.	Eucalyptol.	Quebracho.
Buchu.	Hydronaphthol.	Salol.
Cajuputi.	Kino.	Sassafras.
Catechu.	Matico.	Sumbul.
Cusso.	Menthol.	Thymol.
Diachylon.	Naphthol.	Ichthyol.
Digitalis.	Phenol.	Iodol, etc.

Although we practically always express the quantities by means of the **Roman numerals**, should they be written out they are always in the accusative, formed as follows :

1. Nouns of quantity ending in **a** are feminine and have the accusative singular end in **am** ; plural, **as** ; thus, *drachma*, accusative singular *drachmam;* plural, *drachmas.   Uncia*, accusative singular *unciam;* plural, *uncias.*

2. Those ending in **us** or **um** have the accusative singular end in **um** ; accusative plural of **us** is **os**, and of **um** is **a** ; thus :

	Accusative singular.	*Accusative plural.*
Congius,	Congium,	Congios.
Minimum,	Minimum,	Minima.

Adjectives are declined like nouns and agree with them in gender, number, and case.

If the adjective is of the same declension as the

noun, naturally the genitive will be the same.   If, however, it is of a different declension, its genitive must be formed according to that declension.

To illustrate : Suppose we desire to change the following recipe into unabbreviated Latin—

Take of chlorid of ammonium,   three drams ;
  "   " compound mixture of licorice, two fluidounces.
Mix, and label teaspoonful three times a day.

                                        THOS. LEIDY, M. D.
For JOHN OWENS, *Nov. 15, 1896.*

For " take of " we write ℞. In Latin, chlorid of ammonium is termed *ammonii chloridum ;* but the directions are to take three drams *of* chlorid of ammonium. Consequently, it must be written in the genitive, *ammonii chloridi.*   The accusative plural of *dram* is *drachmas ;* the accusative plural of three is *tres ;* hence, *drachmas tres.*   Compound mixture of licorice in Latin is termed *mistura glycyrrhizæ composita ;* but the direction states *take of* the compound mixture of licorice.   We already have *of licorice* in the phrase ; hence, *glycyrrhizæ* is already in the genitive, but *mistura* and *composita* are not ; they are, according to our table, of the first declension, and consequently form the genitive by changing a to æ ; hence, *misturæ glycyrrhizæ compositæ.*   Then the accusative plural of *uncia* would be *uncias ;* that of *duo, duas—fluiduncias duas.*   For mix we write *misce ;* for label,

5

signa; for three times a day, *t. i. d.*   Therefore, in Latin the prescription would read :

℞ .   Ammonii chloridi,   drachmas tres.
    Misturæ glycyrrhizæ compositæ, fluiduncias duas.
Misce.   Signa.—Teaspoonful, t. i. d.

                                        THOS. LEIDY, M. D.
For JOHN OWENS, *Nov. 15, 1896.*

NOTE.—The salts of the metals generally ending in as change to atis in the genitive, thus :

Nominative.	*Genitive.*
Acetas, ...	Acetatis.
Bicarbonas, ...	Bicarbonatis.
Carbonas, ...	Carbonatis.
Citras, ...	Citratis.
Phosphas, ...	Phosphatis.
Sulphas, ...	Sulphatis.
Tartras, ...	Tartratis, etc.

CHAPTER VI.

DIRECTIONS TO THE APOTHECARY.—LATIN PHRASES AND THEIR ABBREVIATIONS.— NUMERALS.

As we desire to give brief *directions to the apothecary*, there have come into use a number of Latin phrases and words which, however, are generally abbreviated as follows:

Latin.	Abbreviations.	English.
Acidum,	acid.,	an acid.
Ad,	ad,	to, up to.
Ad libitum,	ad lib.,	at pleasure.
Adde,	add.,	add (thou).
Ana,	A., aa,	of each.
Aqua bulliens,	aq. bul.,	water, boiling.
Aqua destillata,	aq. dest.,	water, distilled.
Bene,		well.
Bis in dies,	bis in d.,	twice a day.
Bulliat,	bull.,	boil.
Cape, capiat,	cap.,	take.
Capsula,	caps.,	capsule.
Ceratum,	cerat.,	a cerate.
Chartula,	chart.,	a paper for powder.
Cochleare magnum,	coch. mag.,	tablespoon.
Cochleare parvum,	coch. parv.,	teaspoon.
Cola,	col.,	strain.
Collyrium,	collyr.,	eye-wash.

Latin.	*Abbreviations.*	*English.*
Collutorium,	collut.,	mouth-wash.
Compositus,	comp., co.,	compound.
Congius,	C,	gallon.
Confectio,	conf.,	confection.
Cortex,	cort.,	bark.
Cum,		with.
Decoctum,	decoc.,	a decoction.
Dimidius,.	dim ,	one-half.
Divide,	d., div.,	divide (thou).
Dividendus,	dividend.,	to be divided.
Dividatur in partes æquales,	d. in p. æ.,	divide in equal parts.
Emplastrum,	emp.,	a plaster.
Extende supra,	ex. sup.,	spread upon.
Fac, fiat,	F.,	make, let be made.
Fiant,	F.,	let them be made.
Filtrum, filtra,	fil.,	a filter, filter (v).
Gargarisma,	garg.,	a gargle.
Gutta, guttæ,	gtt.,	drop, drops.
Guttatim,	guttat.,	drop by drop.
Haustus,	haust.,	a draught.
Hora,	h., hor ,	an hour.
In dies,	in d.,	daily.
Instar,	inst.,	like (used with genitive).
Lac,		milk.
Lagena,		a bottle.
Libra,	℔., Lb.,	pound (Troy).
Liquor,	liq.,	solution.
Lotio,	lot.,	lotion.
Massa,	mass.,	pill mass.
Mica panis,	mic. pan.,	crumb of bread.
Misce,	M.,	mix.
Mucilago,	mucil.,	a mucilage.

Latin.	Abbreviations.	English.
Numerus, numero,	No.,	number, in number.
Octarius,	O,	pint.
Ovum, ovi,	ov.,	egg.
Pars,	par.,	a part of.
Partes æquales,	p. æ.,	equal parts.
Pediluvium,		foot-bath.
Per fistulam vitream,		through a glass tube.
Phila,	phil.,	a vial.
Pilula,	pil.,	a pill.
Pro re nata,	p. r. n.,	as required.
Pulvis,	pulv.,	a powder.
Quantum sufficiat,	q. s.,	sufficient quantity of (use genitive after).
Quâquâ horâ,	q. h.,	every hour.
Saturatus,	sat.,	saturated.
Scatula,	scat.,	a box.
Semissis, semisse,	ss.,	one-half.
Sesuncia,	sesunc.,	an ounce and a half.
Signa,	S., Sig.,	sign, label.
Sine,		without.
Solve,	solv.,	dissolve.
Statim,	stat.,	at once.
Talis,	tal.,	such.
Tritura,	trit.,	triturate.
Tere simul,	ter. sim.,	rub together.
Tere exactissime,	ter. exact.,	rub finely.
Ter in die,	t. i. d.,	three times a day.
Vehiculum,	vehic.,	vehicle.
Vitellus,	vit,	yolk of egg.

Although it is perfectly allowable to use the above abbreviations, the names of drugs should never be abbreviated if there is any likelihood of one drug

being taken for another. In fact, it is better always to write out the names of the drugs in full, as the many disastrous results which have already been caused by abbreviating have no need of increase.

Thus, **acid. hydroc.**, may mean either hydrochloric or hydrocyanic acid; **ext. col.**, may mean extractum colchici or extractum colocynthidis; **hydra. chlor.**, either chloral hydrate or chlorid of mercury (hydrargyrum), etc.

NUMERALS

Are all indeclinable except *unus, duo, tres, mille,* and the hundreds (100, 200, etc.).

1, unus.	20, viginti.
2, duo.	30, triginta.
3, tres.	40, quadraginta.
4, quatuor.	50, quinquaginta.
5, quinque.	60, sexaginta.
6, sex.	70, septuaginta.
7, septem.	80, octoginta.
8, octo.	90, nonaginta.
9, novem.	100, centum.
10, decem.	200, ducenti.
11, undecim.	300, trecenti.
12, duodecim.	400, quadringenti.
13, tredecim.	500, quingenti.
14, quatuordecim.	600, sexcenti.
15, quindecim.	700, septingenti.
16, sexdecim.	800, octingenti.
17, septendecim.	900, nongenti.
18, octodecim.	1,000, mille.
19, novendecim.	10,000, decemmille.

Unum, duo, and tres are thus declined :

Masculine.	*Feminine.*	*Neuter.*
Nom., unus,	una,	unum.
Gen., unius,	unius,	unius.
Acc., unum,	unam,	unum.
Nom., duo,	duæ,	duo.
Gen, duorum,	duarum,	duorum.
Acc., duos,	duas,	duo.
Nom., tres,	tres,	tria.
Gen., trium,	trium,	trium.
Acc., tres,	tres,	tria.

CHAPTER VII.

ADMINISTRATION.

Medicines may be given :

1. **By enema** (per rectum, dose about one-third larger than by the mouth).

2. **By hypodermic injection** (subcutaneously, dose about one-third less than by the mouth).

· 3. **By the mouth** (this method employs the officinal doses).

4. **Epidermically** (by means of friction of the skin after local application).

5. **Endermically** (after denudation of the skin by a blister).

6. **Enepidermically** (direct application to skin).

7. **Insufflation** and **vaporization** (usually to respiratory tract only).

8. **Intravenously** (rarely used).

As a rule, the doses stated are intended for adults. In order to employ the same remedies for children several rules have been proposed, viz. :

1. *De Young's Rule.*—Divide the age by the age +· 12, and divide the adult dose by the result.

2. *Cowling's Rule.*—Divide the age at next birthday by 24.

71

3. *Clark's Rule.*—This rule assumes the average weight of a normal adult to be about 150 pounds. To find the dose for a person weighing less than that, divide the dose by the weight divided by 150. Thus: For a child weighing 25 pounds, the dose would be $\frac{25}{150}$, or $\frac{1}{6}$ of the adult dose. This method, although on a more scientific base than the others, is not often used.

To illustrate **De Young's** and **Cowling's** rules: Supposing the adult dose of a drug to be one dram; for a child three years of age we would divide 3 by $3 + 12 = \frac{3}{15}$ or $\frac{1}{5}$; hence, this dose would be 12 grains. According to Cowling's rule, we divide 4 by $24 = \frac{1}{6}$, which equals 10 grains in the above case.

These results are accurate enough for use in all those drugs which are normally well borne by children; it must be borne in mind, however, that some drugs are very poorly borne by children,—for example, morphine, etc.,—while of others,—for example, mercury,—children will often bear well amounts disproportionate to their age and weight.

Doses will also be modified by the general nutrition and vital resistance of the patient, sex (males bear larger doses than females), previous habits (alcoholic, opium, cocain, etc.), previous mode of living, idiosyncrasies, etc.

Some drugs may be taken for long periods of time, but with others the system becomes tolerant and

6

the dose must be increased to have the same, if any, effect.

Again, other drugs—for example, digitalis—have what is called a cumulative effect; *i.e.*, when given for too long a period of time the system seems to store up the drug and there may suddenly be a poisonous explosion, so to speak, in which the poisonous character of the drug may become disagreeably or even fatally manifest.

CHAPTER VIII.

COMBINATION OF MEDICINES.

Different medicines are combined in one mixture in order to (1) increase the action of the base; (2) correct any undesirable action of the base; (3) to meet more than one indication; (4) to obtain the combined effect of the ingredients; (5) to add to the ease of administration; (6) to form new compounds.

The first, second, fourth, and fifth conditions explain themselves.

To illustrate the third : Suppose a patient suffers at the same time with malaria and a cough or cold. By combining, say, quinine and a suitable expectorant, we may control both conditions with one remedy.

To illustrate the sixth : We use daily the combination of potassium iodid and bichlorid of mercury, this giving us the red iodid of mercury, etc.

Under this head, however, we must be careful to employ no remedies which are incompatible with one another.

By incompatibles we may mean any of three classes of compounds, viz. :

1. Unsightly or poisonous mixtures (PHARMACEUTIC INCOMPATIBILITY).

77

2. The formation of new compounds, or the decomposition of the ingredients, which may cause the remedy to become inert, poisonous, or explosive (CHEMIC INCOMPATIBILITY).

3. The combination of remedies having opposite therapeutic effects (PHYSIOLOGIC INCOMPATIBILITY).

Incompatibles of the first class include combinations of oils, balsams, resins (or their alcoholic solutions), tinctures (alcoholic or hydro-alcoholic), fluid extracts (alcoholic), with aqueous solutions, or *vice versâ*.

The second class includes combinations of alkalies (hydrates and carbonates), with acids, oxidizers (peroxids and peracids), with tannin, sugar, sulphur, starch, and sulphids, glycerin, alcohols, and ether; as, for instance, potassium chlorate, or potassium permanganate, with any of the above substances.

Metallic salts with acids, causing precipitation.

Mineral acids with carbonates, acetates, citrates, and salts of the vegetable acids generally.

All alkaloids and their salts are incompatible with tannic acid and all preparations containing it.

Alkaloid salts are incompatible with the alkalies and many of their salts.

Fixed oils and oleoresins may only be employed with water in the form of an emulsion.

Essential oils are soluble in water only to the extent of $\mathfrak{m}$j to f$\mathfrak{Z}$j.

Physiologic incompatibility would include those

compounds containing drugs having opposite actions; as, for instance, a large amount of caffein to keep the patient awake and a correspondingly large amount of opium to put him to sleep.

The following combinations, which might be desirable from a therapeutic standpoint, are incompatible:

Ammonium carbonate, with *syrup of squill,* (which contains *acetic acid*). The *salts* of the heavy metals, *iron, mercury, magnesium,* etc., and the *alkaline earths, lime,* etc., are incompatible with *arsenic.* For example: **Lime-water,** with the *tincture of chlorid of iron,* solutions of *mercurial salts,* etc. **Liquor potassium arsenitis,** with *lime-water,* etc.

Belladonna and Atropin.—*Alkalies* precipitate atropin from solutions of belladonna. *Tannic acid* forms an insoluble tannate with atropin.

Iodin and the iodids are incompatible with *acids* and their salts, and form a heavy insoluble iodid with most alkaloids, which settles to the bottom of the bottle, and so there may be taken in one dose the amount intended for a number of doses. Also incompatible with *soluble metallic salts.*

Iron, or its preparations, form insoluble precipitates with **tannic acid** in any form. Hence we can use only the simple bitters and never any of the vegetable preparations containing tannic acid in combination therewith.

82 FOR THERAPEUTIC NOTES.

Mercury.—**Calomel** is converted into corrosive sublimate when combined with hydrochloric or nitro-hydrochloric acid, the chlorids, and most of the coal-tars (phenacetin, antipyrin, etc.). *Corrosive sublimate* is extremely easily decomposed, and *iodin* changes green mercurous iodid into the more active red iodid.

Nux Vomica.—**Strychnine** in solution *forms* with potassium iodid a precipitate which carries all the strychnia to the bottom of the bottle, rendering one liable to take the whole amount in one dose. (Death has been caused by this combination.)

Opium, Morphine.—**Alkalies** precipitate morphin from solutions of opium. **Tannic acid** forms the very slowly soluble tannate of morphin.

Pepsin, Ingluvin, etc.—**Alcohol** destroys their active properties, as do **alkalies.** **Acids** aid their action if diluted.

Quinine, cinchona.—**Alkalies** precipitate alkaloids of cinchona. *Tannic, gallic, and tartaric acids* form insoluble compounds with them.

Tannic acid and preparations containing it are incompatible with all *alkaloids* and preparations containing alkaloids; with *iron* and preparations containing it.

PART II.

CHAPTER I.

MATERIA MEDICA.

Pharmacology is sometimes defined to mean the action of drugs on the tissues of the living organism; *i. e.*, physiologic action. It is also sometimes supposed to include materia medica, pharmacy, and therapeutics.

Materia medica treats of all the substances, either natural or artificial, used in the practice of medicine. The University of Pennsylvania requires:

1. English name.
2. Scientific name.
3. Physical characteristics.
4. Chemic constituents.
5. Incompatibles.
6. Antidotes.
7. Preparations, United States Pharmacopeia.
8. Doses (crude drug and its preparations).
9. Adulterations, etc.
10. Habitat.

Pharmacy is the science of preparing, compounding, and dispensing medicines.

Therapeutics is the study of the action of medicines in health and disease.

A **pharmacopeia** contains explicit directions for the preparation of medicines, insuring uniformity of strength, action, and nomenclature. It may be officinal (United States) or official. An official pharmacopeia is one sanctioned by the Government; whereas, in the United States it is not sanctioned by the Government but is revised every ten years by representatives of the various recognized schools of medicine, colleges of pharmacy, of the Army and Navy Medical Corps, and of the Marine Hospital Service. Such a pharmacopeia and its preparations are called officinal.

The pharmacopeia does not recognize drugs published under a private formula or whose manufacture is restricted to one or more firms.

The **dispensatory** is much more explicit, giving the physical and chemic history, mode of preparation, doses, and therapeutics of the various drugs, and includes all remedies of whatever nature intended to cure disease.

The principal dispensatories are the "United States Dispensatory" and the "National Dispensatory."

OFFICINAL PREPARATIONS.

Decocta (decoctions) are made by *boiling* the crude drug in water. Useless if the active principle is decomposed by heat or is volatile. If starch is present in any amount it would decompose and thus be useless.

The United States Pharmacopeia recognizes two:

Decoctum—
    Cetrariæ.

Decoctum—
    Sarsaparillæ compositum.

Infusa (infusions) are made by *macerating*, percolating, or displacing the drug in *hot* or *cold water without* boiling. Officinal, four:

Infusum—
    Cinchonæ.
    Digitalis.

Infusum—
    Pruni virginianæ.
    Sennæ compositum.

Liquores (solutions) are solutions of active *non-volatile* principles in *water*. Officinal, 24:

Liquor—
    Acidi arsenosi.
    Ammonii acetatis.
    Arseni et hydrargyri iodidi.
    Calcis.
    Ferri acetatis.
    Ferri chloridi.
    Ferri citratis.
    Ferri et ammonii acetatis.
    Ferri nitratis.
    Ferri subsulphatis.
    Ferri tersulphatis.
    Hydrargyri nitratis.

Liquor—
    Iodi compositus.
    Magnesii citratis.
    Plumbi subacetatis.
    Plumbi subacetatis dilutus.
    Potassæ.
    Potassii arsenitis.
    Potassii citratis.
    Sodæ.
    Sodæ chloratæ.
    Sodii arsenatis.
    Sodii silicatis.
    Zinci chloridi.

Aquæ (waters) are solutions of *volatile* principles in water. Official, 18: *they are all arom ~~thields~~ except the ~~2 anised~~*

Aqua—

Ammoniæ. *10% nat*
Ammoniæ fortior. *28% nat*
Amygdalæ amaræ.
Anisi.
Aurantii florum.
Aurantii florum fortior.
Camphoræ.
Chlori. *.4% L*
Chloroformi.

Aqua—

Cinnamomi.
Creosoti.
Destillata.
Fœniculi.
Hydrogenii dioxidi. *3% Vol*
Menthæ piperitæ.
Menthæ viridis.
Rosæ.
Rosæ fortior.

Spiritus (spirits) are *alcoholic* solutions of *volatile* principles. Official, 25:

Spiritus—

Ætheris.
Ætheris compositus.
Ætheris nitrosi.
Ammoniæ.
Ammoniæ aromaticus.
Amygdalæ amaræ.
Anisi.
Aurantii.
Aurantii compositus.
Camphoræ.
Chloroformi.
Cinnamomi.
Frumenti.

Spiritus—

Gaultheriæ.
Glonoini.
Juniperi.
Juniperi compositus.
Lavandulæ.
Limonis.
Menthæ piperitæ.
Menthæ viridis.
Myrciæ.
Myristicæ.
Phosphori.
Vini Gallici.

Tinctura (tinctures) are *alcoholic* solutions of *non-volatile* principles. Official, 72:

Tinctura—	Per Cent.	Tinctura—	Per Cent.
Aconiti,	35	Aloes et myrrhæ,	10
Aloes,	10	Arnicæ florum,	20

Tinctura— *Per Cent.*

Arnicæ radicis, ... 10
Asafœtidæ, ... 20
Aurantii amari, ... 20
Aurantii dulcis, ... 20
Belladonnæ (Pharm.,
1880), ... 15
Belladonnæ foliorum, · 15
Benzoini, ... 20
Benzoini composita, . 12
Bryoniæ, ... 10
Calendulæ, ... 20
Calumbæ, ... 10
Cannabis indicæ, . . 15
Cantharidis, ... 5
Capsici, ... 5
Cardamomi, ... 10
Cardamomi composita, . 2
Catechu composita, . . 10
Chiratæ, ... 10
Cimicifugæ, ... 20
Cinchonæ, ... 20
Cinchonæ composita, . 10
Cinnamomi, ... 10
Colchici (Pharm., 1880), 15
Colchici seminis, . . 15
Croci, ... 10
Cubebæ, ... 20
Digitalis, ... 15
Tincturæ herbarum recenti-
um, ... 50
Tinctura—
Ferri chloridi, . . . 25
Gallæ, ... 20

Tinctura— *Per Cent.*

Gelsemii, ... 15
Gentianæ composita, . 10
Guaiaci, ... 20
Guaiaci ammoniata, . . 20
Humuli, ... 20
Hydrastis, ... 20
Hyoscyami, ... 15
Iodi, ... 7
Ipecacuanhæ et opii, 100
Kino, ... 10
Krameriæ, ... 20
Lactucarii, ... 50
Lavandulæ composita, . 8
Lobeliæ, ... 20
Matico, ... 10
Moschi, ... 5
Myrrbæ, ... 20
Nucis vomicæ, ... 2
Opii, ... 10
Opii camphorata, less
than . . . . . . ½
Opii deodorati, . . . 10
Physostigmatis, . . . 15
Pyrethri, ... 20
Quassiæ, ... 10
Quillajæ, ... 20
Rhei, ... 10
Rhei aromatica, . . 20
Rhei dulcis, ... 10
Sanguinariæ, ... 15
Saponis viridis (Pharm,
1880), ...
Scillæ, ... 15

94 FOR THERAPEUTIC NOTES.

Tinctura—	Per Cent.	Tinctura—	Per Cent.
Serpentariæ,	10	Valerianæ,	20
Stramonii(Pharm.,1880),	15	Valerianæ ammoniata,	20
Stramonii seminis,	15	Vanillæ,	10
Strophanthi,	5	Veratri viridis,	40
Sumbul,	10	Zingiberis,	20
Tolutana,	10		

The misturæ (mixtures) are simply preparations containing an insoluble substance which is held in suspension in water, usually by the aid of some viscid material. These form in some cases practically an emulsion, and some of them are so called by the 1890 United States Pharmacopeia; whereas in 1880 they were-called mixtures; viz., the 1890 Pharmacopeia recognizes five mixtures:

Mistura—	Mistura—
Cretæ. ✔	Potassii citratis.
Ferri composita. ✔	Rhei et sodæ.
Glycyrrhizæ composita.	

In the 1880 Pharmacopeia the following were called mixtures, but the 1890 Pharmacopeia defines them as EMULSIONS. Official, four:

Emulsum—	Emulsum—
Ammoniaci.	Asafœtidæ.
Amygdalæ.	Chloroformi.

Mucilagines (mucilages) are watery solutions of gummy substances. There are four official mucilages:

Mucilago—	Mucilago—
Acaciæ.	Tragacanthæ.
Sassafras medullæ.	Ulmi.

Syrupi (syrups) are mostly watery solutions of sugary substances, though a few contain dilute acetic acid. Officinal, 31 :

Syrupus—
  Acaciæ.
  Acidi citrici.
  Acidi hydriodici.
  Allii.
  Althææ.
  Amygdalæ.
  Aurantii.
  Aurantii florum.
  Calcli lactophosphatis.
  Calcis.
  Ferri iodidi.
  Ferri, quininæ, et strych-
    ninæ phosphatum.
  Hypophosphitum.
  Hypophosphitum cum ferro.
  Ipecacuanhæ.

Syrupus—
  Krameriæ.
  Lactucarii.
  Picis liquidæ.
  Pruni virginianæ.
  Rhei.
  Rhei aromaticus.
  Rosæ.
  Rubi.
  Rubi idæi.
  Sarsaparillæ compositus.
  Scillæ.
  Scillæ compositus.
  Senegæ.
  Sennæ.
  Tolutanus.
  Zingiberis.

Mellita (honeys) have for their basis the honey of the ordinary honey-bee (*Apis mellifica*). Officinal, two :

Mel despumatum.
  Rosæ.

Aceta (vinegars) are preparations whose menstruum is dilute acetic acid or vinegar. Officinal, two :

Acetum opii.
  Scillæ.

Vina (wines), menstruum of which is white wine, containing 20 to 25 per cent. alcohol. Officinal, ten :

Vinum—
  Album.
  Antimonii.
  Colchici radicis.
  Colchici seminis.
  Ergotæ.

Vinum—
  Ferri amarum.
  Ferri citratis.
  Ipecacuanhæ.
  Opii.
  Rubrum.

Glycerita (glycerites) are solutions in glycerin. Officinal, six :

Glyceritum—
  Acidi carbolici.
  Acidi tannici.
  Amyli.

Glyceritum—
  Boroglycerini.
  Hydrastis.
  Vitelli.

Olei (oils) are volatile or non-volatile, obtained, as a rule, by distillation of plants, although a few are obtained by expression. Officinal, 50 :

Oleum—
  Adipis.
  Æthereum.
  Amygdalæ amaræ.
  Amygdalæ expressum.
  Anisi.
  Aurantii corticis.
  Aurantii florum.
  Bergamii (Pharm., 1880).
  Bergamottæ.
  Betulæ volatile.
  Cadinum (*Juniperi empy-*
  Cajuputi. [*reumaticum*).
  Cari.

Oleum—
  Caryophylli.
  Chenopodii.
  Cinnamomi.
  Copaibæ.
  Coriandri.
  Cubebæ.
  Erigerontis.
  Eucalypti.
  Fœniculi.
  Gaultheriæ.
  Gossypii seminis.
  Hedeomæ.
  Jecoris aselli (*Morrhuæ*).

Oleum—
- Juniperi.
- Juniperi empyreumaticum (*Cadinum*).
- Lavandulæ florum.
- Limonis.
- Lini.
- Menthæ piperitæ.
- Menthæ viridis.
- Morrhuæ (*Jecoris aselli*).
- Myrciæ.
- Myristicæ.
- Olivæ.
- Phosphoratum.
- Picis liquidæ.
- Pimentæ.

Oleum—
- Ricini.
- Rosæ.
- Rosmarini.
- Sabinæ.
- Santali.
- Sassafras.
- Sesami.
- Sinapis volatile.
- Terebinthinæ.
- Terebinthinæ rectificatum.
- Theobromatis (Theobromæ, Pharm., 1880).
- Thymi.
- Tiglii.

Test for Volatile or Non-volatile Oils.—

The volatile oils evaporate entirely if a drop is placed on paper, leaving no greasy mark as do the fixed, non-volatile oils.

Oleoresinæ (oleoresins) are ethereal extracts of drugs containing an oil and a resin. Officinal, six:

Oleoresina—
- Aspidii.
- Capsici.
- Cubebæ.

Oleoresina—
- Lupulini.
- Piperis.
- Zingiberis.

Elixirs (indeclinable) are usually made with a menstruum of dilute alcohol, as are also the succi, or fruit juices (which are juices of the fresh fruits with enough alcohol to preserve them), as succus limonis. Two elixirs are officinal:

Elixir aromaticum.
- Phosphori.

Resinæ (resins) are made by adding water to the saturated alcoholic solutions of the resins, thus causing a precipitation. They are soluble in alcohol but not in water. Officinal, five:

Resina—
    (Residue of turpentine.)
    Copaibæ.
    Jalapæ.

Resina—
    Podophylli.
    Scammonii.

Confectiones (confections) are remedies incorporated into the form of candy. Officinal, two:

Confectio rosæ.
    Sennæ.

Trochisci (troches, lozenges, pellets) are small masses of various shapes made to dissolve slowly in the mouth, mainly to medicate the throat and mouth. Officinal, 15:

Trochisci—
    Acidi tannici.
    Ammonii chloridi.
    Catechu.
    Cretæ.
    Cubebæ.
    Ferri.
    Glycyrrhizæ et opii.
    Ipecacuanhæ.

Trochisci—
    Krameriæ.
    Menthæ piperitæ.
    Morphinæ et ipecacuanhæ.
    Potassii chloratis.
    Santonini.
    Sodii bicarbonatis.
    Zingiberis.

Suppositoria (suppositories) are conic masses of oil of theobroma (butter of cacao) combined with various medicaments, intended for use in the rectum, vagina, and urethra. Rectal and urethral supposi-

tories should weigh about one gram ; for vaginal use, three grams.   Officinal :

Suppositoria glycerini.

Unguenta (ointments) contain of yellow wax 20 per cent., of lard 80 per cent. ; for external use only. Officinal, 22 :

Unguentum—
- Acidi carbolici.
- Acidi tannici.
- Aquæ rosæ.
- Belladonnæ.
- Chrysarobini.
- Diachylon.
- Gallæ.
- Hydrargyri.
- Hydrargyri ammoniati.
- Hydrargyri nitratis.
- Hydrargyri oxidi flavi.

Unguentum—
- Hydrargyri oxidi rubri.
- Iodi.
- Iodoformi.
- Picis liquidæ.
- Plumbi carbonatis.
- Plumbi iodidi.
- Potassii iodidi.
- Stramonii.
- Sulphuris.
- Veratrinæ.
- Zinci oxidi.

Cerata (cerates) contain of white wax 30 per cent. and of lard 70 per cent.   Officinal, five :

Ceratum—
- Camphoræ.
- Cantharidis.
- Cetacei.

Ceratum—
- Plumbi subacetatis.
- Resinæ.

Extracta (extracts) are made by evaporating solutions of vegetable substances, sometimes the fresh juice, resulting in a soft mass.   Officinal, 33 :

Extractum—
- Aconiti.
- Aloes,
- Arnicæ radicis.

Extractum—
- Belladonnæ foliorum alcoholicum.
- Cannabis indicæ.

8

Extractum—
  Cimicifugæ.
  Cinchonæ.
  Colchici radicis.
  Colocynthidis.
  Colocynthidis compositum.
  Conii.
  Digitalis.
  Ergotæ.
  Euonymi.
  Gentianæ.
  Glycyrrhizæ.
  Glycyrrhizæ purum.
  Hæmatoxyli.
  Hyoscyami.

Extractum—
  Iridis.
  Jalapæ.
  Juglandis.
  Krameriæ.
  Leptandræ.
  Nucis vomicæ.
  Opii.
  Physostigmatis.
  Podophylli.
  Quassiæ.
  Rhei.
  Stramonii seminis.
  Taraxaci.
  Uvæ ursi.

Extracta fluida (fluid extracts) are, as a rule, the most powerful of the liquid preparations, and generally *one minim is equal to one grain of the crude drug.* Officinal, 88:

Extractum—
  Aconiti fluidum.
  Apocyni fluidum.
  Arnicæ radicis fluidum.
  Aromaticum fluidum.
  Asclepiadis fluidum.
  Aspidospermatis fluidum.
  Aurantii amari fluidum.
  Belladonnæ radicis fluidum
    (Belladonnæ fluidum,
    Pharm., 1880.)
  Buchu fluidum.
  Calami fluidum.
  Calumbæ fluidum.

Extractum—
  Cannabis indicæ fluidum.
  Capsici fluidum.
  Castaneæ fluidum.
  Chimaphilæ fluidum.
  Chiratæ fluidum.
  Cimicifugæ fluidum.
  Cinchonæ fluidum.
  Cocæ fluidum.
  Colchici radicis fluidum.
  Colchici seminis fluidum.
  Conii fluidum.
  Convallariæ fluidum.
  Cubebæ fluidum.

Extractum—

Cusso fluidum.
Cypripedii fluidum.
Digitalis fluidum.
Dulcamaræ fluidum.
Ergotæ fluidum.
Eriodictyi fluidum.
Erythroxyli fluidum
  (Pharm., 1880).
Eucalypti fluidum.
Eupatorii fluidum.
Frangulæ fluidum.
Gelsemii fluidum.
Gentianæ fluidum.
Geranii fluidum.
Glycyrrhizæ fluidum.
Gossypii radicis fluidum.
Grindeliæ fluidum.
Guaranæ fluidum.
Hamamelidis fluidum
Hydrastis fluidum.
Hyoscyami fluidum.
Ipecacuanhæ fluidum.
Iridis fluidum.
Krameriæ fluidum.
Lappæ fluidum.
Leptandræ fluidum.
Lobeliæ fluidum.
Lupulini fluidum.
Matico fluidum.
Menispermi fluidum.
Mezerei fluidum.
Nucis vomicæ fluidum.
Pareiræ fluidum.

Extractum—

Phytolaccæ radicis fluidum.
Pilocarpi fluidum.
Podophylli fluidum.
Pruni virginianæ fluidum.
Quassiæ fluidum.
Rhamni purshianæ fluidum.
Rhei fluidum.
Rhois glabræ fluidum.
Rosæ fluidum.
Rubi fluidum.
Rumicis fluidum.
Sabinæ fluidum.
Sanguinariæ fluidum.
Sarsaparillæ fluidum.
Sarsaparillæ fluidum compositum.
Scillæ fluidum.
Scoparii fluidum.
Scutellariæ fluidum.
Senegæ fluidum.
Sennæ fluidum.
Serpentariæ fluidum.
Spigeliæ fluidum.
Stillingiæ fluidum.
Stramonii seminis fluidum (Stramonii fluidum, Pharm., 1880.)
Taraxaci fluidum.
Tritici fluidum.
Uvæ ursi fluidum.
Valerianæ fluidum.
Veratri viridis fluidum.

Extractum—
Viburni opuli fluidum.
Viburni prunifolii fluidum.

Extractum—
Xanthoxyli fluidum.
Zingiberis fluidum.

Emplastra (plasters) are prepared by spreading—by means of heat—on muslin, silk, etc., the material intended for external application. Officinal, 13:

Emplastrum—
Ammoniaci cum hydrargyro.
Arnicæ.
Belladonnæ.
Capsici.
Ferri.
Hydrargyri.

Emplastrum—
Ichthyocollæ.
Opii.
Picis burgundicæ.
Picis cantharidatum.
Plumbi.
Resinæ.
Saponis.

Pilulæ (pills) are small rounded masses to be swallowed whole. They include only such substances as are (1) of bad taste; (2) of small dose; (3) of drugs having slow action; (4) of substances too heavy for suspension in liquids, or insoluble therein. For obvious reasons we can not use in pill form substances—(1) of large dose; (2) when rapid action is desired (as emetics, etc.); (3) corrosive substances; (4) deliquescent salts. Officinal, 15:

Pilulæ—
Aloes.
Aloes et asafœtidæ.
Aloes et ferri.
Aloes et mastiches.
Aloes et myrrhæ.
Antimonii compositæ.
Asafœtidæ.
Catharticæ compositæ.

Pilulæ—
Catharticæ vegetabiles.
Ferri carbonatis.
Ferri iodidi.
Opil.
Phosphori.
Rhei.
Rhei compositæ.

Pulveres (powders) are remedies, always dry substances, dispensed in small papers, each of which contains one dose. It must be borne in mind that deliquescent salts, volatile substances, and substances which liquefy when brought together,—chloral and camphor, for example,—are unfit for use in this manner. Officinal, nine : *Those marked important*

Pulvis—

Antimonialis.
Aromaticus.
Cretæ compositus. ✓
Effervescens compositus. ✓
Glycyrrhizæ compositus.

Pulvis—

Ipecacuanhæ et opii.
Jalapæ compositus (*Purgans*).
Morphinæ compositus.
Rbei compositus.

Chartæ (papers) are medicated papers, as a rule, to be burned in order that the vapor of drug they contain may be inhaled, or for external application. Officinal, two :

Charta potassii nitratis.
Sinapis.

Linimenta (liniments) are intended to be used externally, as a rule, to be applied with friction, and are generally of a soapy or oily consistency. Officinal, nine :

Linimentum—

Ammoniæ.
Belladonnæ.
Calcis.
Camphoræ.
Chloroformi.

Linimentum—

Saponis.
Saponis mollis.
Sinapis compositum.
Terebinthinæ.

Collodium (collodions).—Solutions of (ether) in (gun-cotton) which rapidly evaporate when applied to the skin, forming a translucent film containing the medicament and protecting the part to which it is applied.   Officinal, three:

Collodium cantharidatum.
   Flexile.
   Stypticum.

Beside the above-mentioned officinal preparations we have the following

NON-OFFICIAL PREPARATIONS.

Enema (enema or clyster).—Liquids to be injected per rectum.

Bougia.—Small cylinders of cacao butter mixed with the remedy to be used in uterus and urethra.

Pesoaria.—Vaginal suppositories.

Granulum (granules).—Very small pills containing powerful drugs.

Dragées.—Sugar-coated pills of French origin.

Cachets.—These consist of two depressed discs of flour-paper in the interior of which the remedy—generally given in large dose—is placed.   These become very slippery when placed in the mouth, and large doses of drugs disagreeable to the taste may thus be taken without awakening the repugnance of the patient.

Cachous.—Small, highly perfumed pills, often covered with gold- or silver-foil; used mostly to perfume the breath.

CHAPTER II.

AVERAGE DOSES.

Although, of course, we can not give one definite, absolute dose for any drug or drugs, still, for purposes of study, the following list will be found sufficiently accurate. Naturally, the student understands that the dose must be modified to suit the case in hand, as explained in a previous chapter.

Crude drugs may be given in gr. v–x doses.

Exceptions, poisons (see list under Tinctures and Fluid extracts); dose when used, gr. j, about.

Extracts may be given in doses of gr. j–iij.

Exceptions, poisons (see list under Tinctures and Fluid extracts); dose, gr. ⅛–½.

· **Fluid extracts** may be given in ♏x doses.

Exceptions, POISONS: fluid extracts of *aconite, belladonna, digitalis, squill, stramonium, veratrum viride*, the dose of all these being ♏j–ij; and the fluid extracts of *colchicum seed*, ♏ij–x; *colchicum root*, ♏ij–v; *sanguinaria*, ♏j–v; *nux vomica*, ♏j–v.

Tinctures may be given in doses of ʒss–ij.

Exceptions, POISONS: tinctures of *digitalis, iodine, nux vomica, opium*, and *deodorized tincture of opium*, ♏ij–x; *aconite*, ♏j–iij; *belladonna*, ♏x–xx; *col-*

117

118   FOR THERAPEUTIC NOTES.

chicum, ♏ v–xxx ; *physostigma,* ♏ x–xv ; *squill,* ♏ v–xxx.

Infusions and decoctions may be given in f℥ss–ij doses, except infusion of *digitalis,* f℥j–iv.

Syrups may be given in f℥j–ij doses, except the syrups of *iron* and *compound syrup of squill,* ♏ v–xxx.

Mixtures and emulsions may be given in f℥ss doses.

Oils may be given internally in ♏ ij–v–x doses, except Poisons (*croton oil* and *phosphorated oil,* ♏ j) and *castor* oil and *cod-liver* oil, f℥ss.

Alkaloids may be given in doses of gr. $\frac{1}{80}$–$\frac{1}{80}$–$\frac{1}{10}$, except *aconitine,* gr. $\frac{1}{500}$; *caffeine,* gr. ij–xv; *quinine and its associated alkaloids,* gr. ij–xx; *morphine and its associated alkaloids,* gr. ⅛–¼–½ ; *pelletierine and isopelletierine,* gr. v–x.

NOTE.—An alkaloid (vegetable alkali) is "a vegetable extract capable of uniting with acids to form salts," as does ammonia, etc.

A glucoside is "an organic substance which is resolvable by the presence of acids, or other slight chemical influence, into glucose and some other proximate principle" (Example: Certain varieties of tannic acid form glucose and gallic acid, etc.).

CHAPTER III.

OFFICINAL DRUGS AND PREPARATIONS.— IMPORTANT NON-OFFICINAL PREPARA- TIONS.—DOSES.

The study of materia medica renders necessary the knowledge of the following points in reference to each and every substance employed therein, viz. :

1. English name.
2. Scientific name.
3. Physical characteristics.
4. Chemic constituents.
5. Incompatibles (if any).
6. Antidotes (if poisonous).
7. Preparations (United States Pharmacopeia).
8. Doses (of crude drug and its preparations).
9. Adulterations.
10. Habitat.
11. How made (of chemicals).
12. Parts of plant used (if of vegetable kingdom).
13. Parts of animal used (if animal).

The classification adopted by Dr. H. C. Wood in "Therapeutics : Its Principles and Practice," will be followed in the consideration of the drugs to follow :

Division I.—Systemic Remedies.

Division II —Extraneous Remedies.

SYSTEMIC REMEDIES

May again be divided into—

(*a*) *General remedies*, "affecting the tissues of the body generally, or such organized systems as reach all portions of the body."

(*b*) *Local remedies*, those affecting only one organ of the body.

The general remedies may be divided into the following three orders:

I. Nervines, acting on the nervous system.

II. Cardiants, acting on the circulation.

III. Nutrients, acting on the general nutrition.

NERVINES.

A. Medicines acting on the cerebrum.

B. Medicines acting on the remainder of the nervous system.

A. FAMILY 1.—**Antispasmodics**, employed for the relief of minor spasms and nervous manifestations, are feeble cerebral stimulants.

FAMILY 2.—**Anesthetics** are used to produce anesthesia (local and general).

FAMILY 3 —**Somnifacients**, in proper doses produce sleep without delirium.

FAMILY 4 —**Delirifacients**, in proper doses produce delirium first, then stupor.

B. FAMILY 5.—**Excito-motors**—drugs causing violent tetanic spasms in overdose.

FAMILY 6.—Depresso-motors — drugs causing paralysis in overdose.

CARDIANTS.

FAMILY 1.—Cardiac stimulants—drugs which increase arterial pressure.

FAMILY 2.—Cardiac depressants—drugs which decrease arterial pressure.

NUTRIENTS.

FAMILY 1.—Astringents—drugs causing contraction of various organs.

FAMILY 2.—Tonics—drugs which increase nutrition and vital power.

FAMILY 3.—Alteratives—drugs which modify nutrition and overcome " certain chronic pathologic processes."

FAMILY 4.—Antiperiodics—drugs which overcome the effects of malarial poisoning.

FAMILY 5.—Antipyretics—drugs which overcome febrile movements.

Local Remedies.

FAMILY 1.—Stomachics.
FAMILY 2 —Emetics.
FAMILY 3.—Cathartics.
FAMILY 4.—Diuretics.
FAMILY 5.—Diaphoretics.
FAMILY 6.—Expectorants.
FAMILY 7.—Emmenagogues.
FAMILY 8.—Oxytocics.
FAMILY 9.—Sialagogues.
FAMILY 10.—Errhines.
FAMILY 11.—Epispastics.
FAMILY 12.—Rubefacients.
FAMILY 13.—Escharotics.
FAMILY 14 —Demulcents.
FAMILY 15.—Emollients.
FAMILY 16.—Protectives.

EXTRANEOUS REMEDIES.

FAMILY 1.—Antacids.
FAMILY 2.—Anthelmintics.
FAMILY 3.—Digestants.
FAMILY 4.—Absorbents.
FAMILY 5.—Disinfectants.

128   FOR THERAPEUTIC NOTES.

ORDER I.—NERVINES.

FAMILY I.—ANTISPASMODICS.

Drugs capable of controlling minor spasms of voluntary or involuntary muscles.

Officinal Name, MOSCHUS. *English Name*, MUSK.

Definition.—Dried *secretions* from the *preputial follicles* of Moschus moschiferus (musk deer).

Class.—Mammalia. *Order.*—Ruminantia. *Habitat.* — Thibet. *Physical Properties.* — Irregular, crumbly grains; dark red-brown; peculiar, penetrating, persistent odor; bitter taste.

The crude drug is given in **doses** of gr. v–xv in emulsion, capsule, or injection (per rectum).

Officinal Preparation.

Tinctura Moschi, ... f ℥ ss–ij.

Officinal Name, VALERIANA. *English Name*, VALERIAN.

Definition.—The *rhizome* and *rootlets* of Valeriana officinalis.

Natural Order.—Valerianeæ. *Habitat.*—Europe. *Physical Properties.*—Its peculiar odor, resembling perspiration (due to *valerianic acid*), becomes stronger on keeping; taste camphoraceous and bitter. Contains *valerianic acid* and *oil of valerian*, both volatile.

Officinal Preparations.

Extractum Valerianæ Fluidum, . . f ℥ ss–j.
Tinctura Valerianæ, ... f ℥ ss–ij.
Tinctura Valerianæ Ammoniata, . . f ℥ ss–ij.
Ammonii Valerianas, ... gr. ij–x.

Iron, quinine, and *zinc valerianate* will be considered under iron, quinine, and zinc.

✔ *Officinal Name,* ASAFŒTIDA. *English Name,* ✔
ASAFETIDA.'

Definition.—A *gum-resin* obtained from the *root* of Ferula fœtida.

Natural Order.—Umbelliferæ. *Habitat.*—Afghanistan. *Physical Properties.*—Irregular masses of whitish tears imbedded in a yellowish-gray, sticky mass; has a persistent, alliaceous odor, and bitter, acrid, alliaceous taste. Contains a volatile oil, a gum, and a resin.

*Officinal Preparations.**

Emulsum or Mistura (Pharm., 1880)
Asafœtidæ, ... f ℥ ss–j.
Pilulæ Aloes et Asafœtidæ (aloes, asafetida, and soap, of each gr. 1⅓).
Pilulæ Asafœtidæ, each, ... gr. iij.
Tinctura Asafœtidæ, ... f ℥ ss–ij. ✓

✔ *Officinal Name,* CAMPHORA. *English Name,* CAMPHOR.

Definition.—A stearopten (having the nature of a ketone) from the Cinnamomum camphora.

---

* Mistura magnesiæ et asafœtidæ (Dewees' Carminative), no longer officinal, contains mag. carbonate, seven per cent.; laudanum, one per cent.; tinct. asafœtidæ, seven per cent.; sugar and water.

amphor $C_{10} H_{16}$

ficial Camphor $C_{10} H_{16} O$

camphora Monobromata $C_{10} H_{15} Br$

Amber — succina — not a
more official,

Used also as Hoffman
anodyne is the
Spiritus Aetheris.

Natural Order.—Laurineæ. *Habitat.*—China and Japan. *Physical Properties.*—White, tough, translucent masses ; crystalline, penetrating, characteristic odor, and pungent, aromatic taste. Pulverizable in presence of alcohol, ether, or chloroform.

Officinal Preparations.

Aqua Camphoræ (1 to 125), ... f℥ ss–ij.
Linimentum Camphoræ (20 per cent. solution in cotton-seed oil).
Linimentum Saponis.
Spiritus Camphoræ (ten per cent. camphor), ... gtt. v–fℨj.
Tinctura Opii Camphorata (see Opium).
Camphora Monobromata, ... gr. iij–xv.

Officinal Name, SPIRITUS ÆTHERIS COMPOSITUS.
English Name, COMPOUND SPIRIT OF ETHER (HOFFMANN'S ANODYNE).

Definition.—An alcoholic solution of ether and ethereal oil made as follows :

Ether, ... 325 c. c.
Alcohol, ... 650 "
Ethereal Oil, ... 25 "
_____
1000 c. c.

The genuine preparation imparts a cloudiness to water when about 45 drops of it are added to a pint of water. Adulterations remain clear, owing to absence of oil.

Dose, fℨss–ij.

Officinal Name, HUMULUS. *English Name*, HOPS. ✓

Definition.—*Strobiles* of Humulus lupulus.

Natural Order.—Urticaceæ. *Habitat.*—Europe and North America. *Physical Properties.*—Ovate scales containing a glandular, yellowish powder called *lupulin,* to which its activity is due. Scales themselves are greenish, have an aromatic odor, bitter, astringent taste. Crude drug used only for poultices.

Officinal Preparations.

Tinctura Humuli, ... f ℨ ss–ij.
Lupulinum, ... gr. x–xx.
Extractum Lupulini Fluidum, ... f ℨ ss–j.
Oleoresina Lupulini, ... ℳ v–xxx.

Officinal Name, LACTUCARIUM. *English Name*, LACTU-
CARIUM, LETTUCE OPIUM. ✓

Definition.—Concrete *milky juice* of Lactuca virosa.

Natural Order.— Compositæ. *Habitat.*—Indigenous.

Dose of crude drug, gr. x–ʒj.

Officinal Preparation.

Tinctura Lactucarii, ... f ℨ ss–f ℥ ij.

Officinal Name, CIMICIFUGA. *English Name*, BLACK
SNAKEROOT, BLACK COHOSH.

Definition.—*Rhizome* and *rootlets* of Cimicifuga racemosa.

Natural Order. — Ranunculaceæ. *Habitat.* — United States.

Oxide of ethyl = true ether
of U.S.P. L $(C_2 H_5)_? O$
Acetic Ether = $C_2 H_5 C_? H_3 O_2$

Extractum Cimicifugæ, ... gr. ¼–iij.
Extractum Cimicifugæ Fluidum, . . . ʄ ℥ ss–j.
Tinctura Cimicifugæ, ... ʄ ℥ ss–ij.

FAMILY II.—ANESTHETICS.

General anesthetics produce loss of consciousness of entire body.

Local anesthetics cause loss of sensation only in the part of the body to which they are applied,—practically only the mucous surfaces and superficial regions of the body.

NITROUS OXID (N_2O), LAUGHING GAS, NITROGEN MONOXID.*

Definition.—Colorless, practically odorless, gas, made by heating ammonium nitrate.

$$NO_3NH_4 + heat = N_2O + 2H_2O.$$

Used mainly in dentistry. Administered by inhalation.

Officinal Name, ÆTHER. *English Name*, ETHER, SULPHURIC ETHER (*incorrect*), ÆTHER FORTIOR (Pharm., 1880).

Definition.—A volatile liquid prepared by distilling alcohol in presence of sulphuric acid. Contains 96 per cent., by weight, of ethyl oxide [$(C_2H_5)_2O$], and about four per cent. of alcohol. Specific gravity, 0.728.

---

* Not officinal.

oform $CH\,Cl_3$ — Trichlor
methaur.

ichlor — methanal $C\,H_2\,Cl_2$.
~~~~ ~~~~ ~~ ?
~~~~~ .

Physical Properties.—Transparent, colorless, mobile liquid, of characteristic odor and sweetish, burning taste. Volatile and highly inflammable. The vapor, if mixed with air and ignited, explodes violently; is heavier than air, and therefore may be used in the presence of artificial light, provided the light is kept higher than the vapor.

Officinal Preparations.

Spiritus Ætheris (30 per cent.), . . . f ʒ ss–ij.
Spiritus Ætheris Compositus, ... f ʒ ss–ij.

Officinal Name, CHLOROFORMUM (CHLOROFORMUM PURIFICATUM, Pharm., 1880). *English Name*, CHLOROFORM.

Definition.—Volatile liquid obtained by distilling alcohol in the presence of chlorinated and slaked lime. It is clear, heavy, and colorless, has a sweet taste and characteristic ethereal odor. Contains 99 per cent., by weight, of absolute chloroform and one per cent. alcohol. Specific gravity, 1.490 (about). It is not inflammable, but its vapor burns with a green flame.

Officinal Preparations.

Aqua Chloroformi, ... f ℥ ss.
Emulsum Chloroformi, ... f ℥ ss.
Linimentum Chloroformi, 40 per cent.
   chloroform in soap liniment.
Spiritus Chloroformi, ten per cent., . f ʒ ss–ij.

Bichloride of methylene, bromoform, and ethyl bromide somewhat resemble chloroform, and are occasionally used as anesthetics.

Local Anesthetics.—Cocaine (see Coca) is used as a local anesthetic on mucous membranes by direct application and consequent absorption, or under skin surfaces by hypodermic injection after checking the blood-supply to and from the part.

Dose of the hydrochlorate, gr. ¼.

Eucaine * is claimed to be less poisonous and dangerous than cocaine. Used in same dose and manner.

The rapid evaporation of chloride of ethyl produces the same effect as freezing of a part, and many small operations can be done under its influence. Contained in glass capsules, from which it is sprayed on the part until anesthesia follows.

FAMILY III.—SOMNIFACIENTS.

Somnifacients are drugs used to produce sleep.

Officinal Name, OPIUM.   *English Name*, OPIUM.

Definition.—The *concrete, milky exudation* obtained by incising the unripe capsules of Papaver somniferum (poppy).

Natural Order.—Papaveraceæ.   *Habitat.*—Asia Minor, Persia, India, etc.   *Physical Properties.*— Irregular cakes wrapped in poppy leaves ; plastic or harder ; chestnut brown or darker ; sharp, narcotic odor, and peculiar, bitter taste.   Must yield in the

---

* Not officinal.

normal moist condition not less than nine per cent. of **morphine,** its principal alkaloid.

Officinal.

Opii, ...
— Opil Pulvis (not less than 13 nor more
  than 15 per cent. of morphine), . } Dose, gr. ẍ-ij.
Opium Deodoratum (Opium Denarco-
  tisatuḿ, Pharm., 1880), ...

Officinal Preparations.

Acetum Opii, ...
Vinum Opii, ...
Tinctura Opii (laudanum), ... } ten per cent. of
Tinctura Opii Deodorati, ... opium. Dose,
Tinctura Ipecac. et Opii (corresponds ♏x-xx.
  to Dover's powder), ...

Tinctura Opil Camphorata (paregoric)
  contains two grains of opium to one
  fluidounce; also benzoic acid, cam- } Dose, f ℥ ss-iv.
  phor, oil of anise, glycerin, and di-
  lute alcohol, ...

Extractum Opii, ... gr. ss.
Pilulæ Opii, each contain of powdered
  opium, ... gr. j.
Pulvis Ipecac. et Opii (*Dover's powder*)
  (O., 1; ip., 1; sacch. lactis, 8), . . gr. x.
Trochisci Opii et Glycyrrhizæ, . . . x = gr. j.
Emplastrum Opii, *ext. opii,* . . . . . 6 in 100.

Opium contains a number of alkaloids in com-bination with meconic and thebolactic acids. Only **morphine** and **codeine** are officinal.

Officinal Preparations.

Morphina, used in pharmacy.

Morphinæ Sulphas, ...⎫

Morphinæ Acetas, ...⎬ gr. ⅛–½.

Morphinæ Hydrochloras, ... ⎭

Pulvis Morphinæ Compositus (morph.
   sulph., 1 ; camphor, 19; excipient,
   80) (*Tully's powder*), ... gr. x.

Trochisci Morphinæ et Ipecacuanhæ, . 1 = gr. $\frac{1}{40}$.

Codeina, ... gr. ¼–ij.

Narceina,* Thebaina or Paramorphina,*
   Narcotina,* Laudanina,* Meconina,*
   Papaverina,* Porphyroxina,* etc., . gr. ij–x.

Apomorphinæ Hydrochloras (see Emetics).

Solutions of opium produce a blood-red color on the addition of ferric chloride (due to meconic acid). Morphine strikes a deep blue with ferric chloride and a rich orange-red, fading into yellow, with concentrated nitric acid. (See different result with quinine, under Cinchona.)

If morphine is treated with cold, concentrated sulphuric acid (free from nitric acid), on subsequent addition of a small crystal of potassium permanganate a greenish color only should be produced. Strychnine gives with this test a violet or purple color.

(For Poisoning see chap. on Antidotes.)

Officinal Name, CHLORAL. *English Name,* CHLORAL
   HYDRATE OR CHLORAL.

Definition.—A volatile, crystalline solid, of aro-

---

* Not officinal.

matic, acrid odor, and bitter, caustic taste. Obtained by acting on alcohol with chlorine.

Preparation.

Chloralis, ... gr. x–xxx.

Chloral forms with camphor, chloral-camphor—a liquid for external use only.

HYOSCINÆ HYDROBROMAS, gr. $\frac{1}{10}$. (See Hyoscyamus.)

METACHLORAL,* CHLORALAMID,* CHLORALOSE,* AND BUTYL CHLORAL HYDRATE,*

All modifications of chloral, are given in the same manner and dose.

SULPHONAL*

Is a valuable somnifacient in doses of gr. x–ʒss.

TRIONAL AND TETRONAL*

Are given in the same dose as sulphonal.

Officinal Name, PARALDEHYDUM. *English Name*, PAR-ALDEHYDE (C_3H_4O).

Definition.—Acetic aldehyde caused by oxidation, as in the effect of chromic acid on alcohol.

Dose, ♏ xx–xl.

FAMILY IV.—DELIRIFACIENTS.

Remedies causing marked dilatation of the pupil, and acting on the cerebrum; causing delirium in overdose.

* Not officinai.

Officinal Name, CANNABIS INDICA. *English Name*,
INDIAN HEMP, INDIAN CANNABIS.

Definition.—The *flowering tops* of the *female*
plant of the Cannabis sativa, grown in the East
Indies.

Natural Order.—Urticaceæ. *Habitat.*—East Indies.

Officinal Preparations.

Extractum Cannabis Indicæ, ... gr. ¼-j.
Extractum Cannabis Indicæ Fluidum, . ♏j-x.
Tinctura Cannabis Indicæ, ... ♏x-xxx.

Officinal Name,   *English Name*,
BELLADONNÆ FOLIA,   BELLADONNA LEAVES,
BELLADONNÆ RADIX.   BELLADONNA ROOT.

Definition.—*Leaves* and *root* of Atropa belladonna (deadly nightshade), a European perennial.

Natural Order.—Solanaceæ. *Habitat.*—Europe.

Officinal Preparations.

Leaf:

Extractum Belladonnæ Foliorum Alcoholicum, ... gr. ⅛-½.
Tinctura Belladonnæ Foliorum, ... ♏x-xxx.

Root:

Extractum Belladonnæ Radicis Fluidum, ♏j-ij.
Emplastrum Belladonnæ, 2 parts of
alcoholic ext. to 10 parts of plaster.
Linimentum Belladonnæ, camphor, 50
parts, fluid extract belladonna to make 1000.

Belladonna contains the alkaloid **atropine**, offici-nal as **atropinæ sulphas**, gr. $\frac{1}{200}-\frac{1}{50}$.

An alcoholic solution of atropine added to mercuric chloride gives a yellow precipitate turning to red.

It may also be tested **physiologically** by applying the suspected solution to an eye of a lower animal. Atropine invariably causes dilatation of the pupil.

Atropine is incompatible with tannic acid, and alkalies precipitate atropine from the various solutions of belladonna.

Homatropine is an artificial alkaloid and is sometimes preferred as a mydriatic, its effect passing off sooner than that of atropine.

(For Poisoning see chap. on Antidotes.)

Officinal Name, STRAMONII SEMEN. *English Name*, STRAMONIUM SEED.

Definition.—The *seeds* of the Datura stramonium, or Jamestown (Jimpson) weed.

Natural Order.—Solanaceæ. *Habitat.*—United States.

Officinal Preparations.

Extractum Stramonii Seminis, . . . gr. ¼–j.
Extractum Stramonii Seminis Fluidum, ♏j–v.
Tinctura Stramonii Seminis, ... ♏v–xx.
Unguentum Stramonii (ten per cent. of extract).

Stramonium yields the alkaloid **daturine**, which is practically the same in action as atropine, but is not officinal.

[handwritten notes, largely illegible]

... formulae of Atropine,
... and ...
... $C_{17}H_{23}$ ...

Officinal Name, HYOSCYAMUS. *English Name*, HEN-
BANE.

Definition.—The *leaves* and *flowering tops* of
Hyoscyamus niger, from plants of the *second year's*
growth.

Natural Order.—Solanaceæ. *Habitat.*—Europe.

Officinal Preparations.

Extractum Hyoscyami, ... gr. j–ij.
Extractum Hyoscyami Fluidum, . . . ℳ v–xx.
Tinctura Hyoscyami, ... f ℨ ss–j.

Hyoscyamus yields the alkaloids hyoscine (hydro-
bromate), gr. $\frac{1}{160}$–$\frac{1}{80}$, and hyoscyamine (sulphate
and hydrobromate), gr. $\frac{1}{80}$; both are officinal.

Officinal Name, COCA (ERYTHROXYLON, Pharm., 1880).
English Name, COCA.

Definition.—The *leaves* of the Erythroxylon
coca.

Natural Order. — Lineæ. *Habitat.* — South
America.

Officinal Preparation.

Extractum Cocæ Fluidum, ... f ℨ ss–ij.

Its alkaloid is cocaine (hydrochlorate), dose,
gr. ¼–j, used mostly as a local anesthetic.

Tropacocaine, from the narrow-leaved coca of
Java, resembles, for all practical purposes, cocaine in
its action.

II

FAMILY V.—EXCITO-MOTORS.

These remedies, by exciting the reflex centers of the spinal cord, produce in normal doses increased muscular activity; in poisonous doses they produce tetanic convulsions.

Officinal Name, NUX VOMICA. *English Name*, NUX VOMICA.

Definition.—The *seed* of Strychnos nux-vomica, a small East Indian tree.

Natural Order.—Loganiaceæ. *Habitat.*—East Indies.

Officinal Preparations.

Extractum Nucis Vomicæ, ... gr. ¼–ss.
Extractum Nucis Vomicæ Fluidum, . . ♏ j–v.
Tinctura Nucis Vomicæ (20 per cent.), ♏ ij–x.

Alkaloids.—Strychnine (officinal) and brucine (non-officinal). Strychninæ sulphas, gr. $\frac{1}{60}-\frac{1}{20}$.

Test.—A crystal of potassium dichromate drawn through a solution of strychnine in concentrated sulphuric acid produces a blue color changing to violet, purple-red, then orange or yellow.

Brucine gives a blood-red color, fading into yellow, with nitric acid.

(For Poisoning see chap. on Antidotes.)

Officinal Name, IGNATIA (Pharm., 1880). *English Name*, IGNATIA:

Definition.—This drug, no longer officinal, is

like nux vomica in action and contains also strychnine and brucine.

FAMILY VI.—DEPRESSO-MOTORS.

By depressing the spinal centers these remedies
lessen muscular activity.

Officinal Name, PHYSOSTIGMA. *English Name*, CALA-
BAR BEAN.

Definition.—The *seed* of Physostigma veneno-
sum.

Natural Order.—Leguminosæ. *Habitat.*—West
coast of Africa.

Contains the alkaloid physostigmine (or eserine).

Officinal Preparations.

Extractum Physostigmatis, ... gr. ⅙–j.
Tinctura Physostigmatis, ... ♏v–xx.

Physostigmine or Eserine, its alkaloid, is
officinal as :

Physostigminæ Salicylas, ... } gr. 1/60–1/10.
Physostigminæ Sulphas, ... }

The bromides belong to this class.

Potassii Bromidum, ... gr. v–ʒj.
Lithii Bromidum, ... gr. x–ʒss.
Sodii Bromidum, ... gr. v–ʒj.
Ammonii Bromidum, ... gr. v–ʒss.
Acidum Hydrobromicum Dilutum (ten
Per cent.), ... ʒss–j.

Officinal Name, AMYL NITRIS. *English Name*, AMYL
NITRITE.

Definition.—A yellow, oily, very volatile liquid,
containing 80 per cent. of *amyl* (mainly isoamyl).
Its odor resembles that of fruit. Almost insoluble
in water ; miscible with alcohol or ether. Prepared
by the action of nitric acid on amylic alcohol (fusel
oil).

Dose, m j–x by inhalation ; m j–v, internally.

Officinal Name, SPIRITUS GLONOINI. *English Name*,
SPIRIT OF NITROGLYCERIN (GLONOIN).

Definition.—An alcoholic (one per cent. by
weight) solution of glonoin (propenyl) trinitrate.
Clear and colorless, resembling alcohol in odor and
taste. Great caution is necessary in handling it,
owing to its extremely explosive properties. Should
the alcohol evaporate explosion may occur. Tasting
even a small amount may cause violent headache.
Prepared by the action of nitric acid on glycerin.

Dose, m ss–j.

Officinal Name, LOBELIA. *English Name*, INDIAN
TOBACCO.

Definition.—The *leaves* and *tops* of Lobelia in-
flata.

Natural Order.—Lobeliaceæ. *Habitat.*—United
States.

Contains the *liquid* alkaloid lobeline (not officinal) and lobelic acid.

Officinal Preparations.

Extractum Lobeliæ Fluidum, ... ♏v–xv.

Tinctura Lobeliæ, ... ♏v–xxx.

Officinal Name, GELSEMIUM. *English Name*, YELLOW, OR CAROLINA, JASMINE.

Definition.—*Rhizome* and *rootlets* of Gelsemium sempervirens.

Natural Order.—Loganiaceæ. *Habitat.*—Southern United States.

Contains the alkaloid **gelsemine**, and gelseminic acid (non-officinal).

Officinal Preparations.

Extractum Gelsemii Fluidum, ... ♏iij–v.

Tinctura Gelsemii, ... ♏x–xxx.

Officinal Name, ACIDUM HYDROBROMICUM DILUTUM. *English Name*, DILUTED HYDROBROMIC ACID.

Definition.—Ten per cent. aqueous solution of absolute hydrobromic acid.

Dose, f℈ss–ij.

Officinal Name, TABACUM. *English Name*, TOBACCO.

Definition.—The commercial, dried *leaves* of Nicotiana tabacum.

Natural Order.—Solanaceæ. *Habitat.*—United States.

Contains the *liquid volatile* alkaloid **nicotine** and a volatile oil. Rarely used. There are no officinal preparations.

Officinal Name, CONIUM. *English Name*, HEMLOCK.

Definition.—The full-grown *fruit* of Conium maculatum or hemlock gathered while yet green. Its activity depends on conine, *a volatile liquid* alkaloid (not officinal).

Natural Order.—Umbelliferæ. *Habitat.*—United States and Europe.

Officinal Preparations.

Extractum Conii, ...	gr. j-ij.
Extractum Conii Fluidum, ...	♏j-v.

As the preparations vary greatly in strength, begin always with the minimum dose and increase to desired effect.

FAMILY VII.—RESPIRATORY STIMULANTS.

Although classified under different headings, ammonia, caffeine, atropine, cocaine, and strychnine are valuable also as respiratory stimulants.

Officinal Name, ASPIDOSPERMA. *English Name*, QUEBRACHO.

Definition.—The *bark* of Aspidosperma quebracho-blancho (Schlechtendal).

Natural Order.—Apocynaceæ. *Habitat.*—South America.

Officinal Preparation.

Extractum Aspidospermatis Fluidum, . f℥ ¼-½.

Contains the alkaloid aspidospermine.
Dose, gr. ¼-ss.

ORDER II.—CARDIANTS.

FAMILY I.—CARDIAC STIMULANTS.

These remedies increase the force of the circulation, either by stimulating the heart muscle directly or by lessening resistance to the flow of blood; *i. e.*, dilating the arteries through which it runs.

AMMONIA.*

Definition.—A colorless irritant gas, alkaline in reaction, characteristic in odor, extremely soluble in water. Naturally found as a result of decaying animal and vegetable matter. Commercially obtained as a by-product in the manufacture of coal-gas.

Officinal Preparations.

Spiritus Ammoniæ, ... ♏v–f℥ss.
Spiritus Ammoniæ Aromaticus, . . f℥ss–ij.
Aqua Ammoniæ, ten per cent. gas, . . ♏x–xx.
Aqua Ammoniæ Fortior, 28 per cent.
  gas, ... ♏ij–x.
Liquor Ammonii Acetatis (dilute acetic
  acid neutralized by carbonate of am-
  monium—*Spirits of Mindererus*), . f℥ss–ij.
Linimentum Ammoniæ.
Ammonii Benzoas, ... }
Ammonii Bromidum, ... } gr. x–xxx.
Ammonii Carbonas, ... gr. ij–x.
Ammonii Chloridum, ... gr. v–℥ss.

---

* Not officinal as gas.

Trochisci Ammonii Chloridi, . . ā ā gr. $\frac{1}{10}$.

Ammonii Iodidum, ... gr. ij–x.

Ammonii Phosphas, ... gr. x–xv.

Ammonii Valerianas, ... gr. ij–x.

Officinal Name, ALCOHOL.

Definition.—A transparent volatile liquid; characteristic odor and taste; 91 per cent., by weight, of ethyl alcohol, and nine per cent., by weight, of water; specific gravity, 0.820.

Officinal Preparations.

Alcohol Dilutum.

Alcohol Absolutum, specific gravity, 0.797; contains only one per cent. water.

Alcohol Deodoratum, specific gravity, 0.816; 7½ per cent. water.

Officinal Name, ALCOHOL DILUTUM. *English Name*, DILUTED ALCOHOL.

Definition.—A liquid composed of 41 per cent., by weight, or about 48.6 per cent., by volume, of absolute ethyl alcohol, and about 59 per cent., by weight, of water. Alcohol exists in the following *officinal preparations*, viz. :

Spiritus Frumenti.—*Whisky*, 44 to 50 per cent. alcohol, by weight. Obtained by the distillation of fermented grains (practically distilled beer). Must be *at least two* years old.

Spiritus Vini Gallici.—*Brandy*, 39 to 47 per cent. alcohol, by weight. Obtained by distillation of fermented grapes (practically distilled wine). Must be *at least four* years old.

italis has a cumulative eff
after 14 days administrate
be discontinued for a

an.

prima excellent
idote for one plum poisoning

Vinum Album.—*White wine*, 10 to 12 per cent. alcohol.

Vinum Rubrum.—*Red wine*, 10 to 12 per cent. alcohol.

Vinum Album Fortius.—Twenty to 25 per cent. alcohol, used as menstruum for officinal wines.

Spiritus Odoratus.—*Cologne water.*

Vinum Aromaticum.—*Aromatic wine.*

Beer and the various malts (extracts, etc.) vary from three to eight per cent. of alcohol.

Officinal Name, DIGITALIS. *English Name*, FOXGLOVE (PURPLE).

Definition.—*Leaves* of Digitalis purpurea from plants of *second year's* growth. Contains four glucosides, viz. : *digitalin, digitoxin, digitonin,* and *digitalein.*

Natural Order.—Scrophularineæ.   *Habitat.*— Europe.

Powdered leaves may be used in **doses** of gr. ss–iij.

Officinal Preparations.

Extractum Digitalis, ... gr. ¼–j.
Extractum Digitalis Fluidum, ... ℳ ss–ij.
Tinctura Digitalis, ... ℳ v–xv.
Infusum Digitalis, ... ʒ j–v.

Officinal Name, CAFFEINA. *English Name*, CAFFEINE.

Definition.—Feeble alkaloid obtained from *dried leaves* of Thea sinensis (*natural order*, Ternstrœmiaceæ), ordinary tea, from Coffea arabica (*natural order*, Rubiaceæ), and from guarana * (a paste from

------

* Officinal.

12

[handwritten note, largely illegible]

tea Arabica contains about
1½ % Caffeine; then
a little more...
taining most — 2½ to 4 %.

crushed seeds of Paullinia sorbilis, *natural order*, Sapindaceæ).

Officinal Preparations.

Caffeina, ... gr. ij–x.
Caffeina Citrata, ... gr. x–xv.
Caffeina Citrata Effervescens, ... gr. x–xv.

Officinal Name, CONVALLARIA. *English Name*, LILY OF THE VALLEY.

Definition.—The *rhizome* and *rootlets* of Convallaria majalis, or lily of the valley.

Natural Order. — Liliaceæ. *Habitat.* — United States, Holland.

Contains the glucosides *convallarin* and *convallamarin*.

Officinal Preparation.

Extractum Convallariæ Fluidum, ... ℳv–xv.

Officinal Name, STROPHANTHUS. *English Name*, STROPHANTHUS.

Definition.—*Seeds* of Strophanthus hispidus deprived of its long awn. Used as arrow-poison by the African natives.

Natural Order.—Apocynaceæ. *Habitat.*—Africa.

Contains the active principle *strophanthin*.

Officinal Preparation.

Tinctura Strophanthi, ... ℳj–x.

Officinal Name, SPARTEINÆ SULPHAS. *English Name,*
SPARTEINE SULPHATE.

Definition.—The neutral sulphate of an alkaloid obtained from Scoparius or broom plant.

Natural Order.—Leguminosæ. *Habitat.*—Indigenous.

Dose, gr. $\frac{1}{20}$–¼ hypodermically; gr. ¼–ij in pill.

ADONIDINE.*

Definition.—A *glucoside* from *root* of Adonis vernalis.

Natural Order. — Ranunculaceæ. *Habitat.* — Europe.

Dose, gr. ⅛–ss.

FAMILY II.—CARDIAC DEPRESSANTS.

These remedies depress the heart's action, and so reduce the force and frequency of the pulse.

ANTIMONIUM.†

Antimony (metallic element) is found as black antimonious sulphide.

Officinal Preparations.

Antimonii et Potassii Tartras (*Tartar Emetic*). Dose, as diaphoretic and expectorant, gr. $\frac{1}{12}$–⅛; as emetic, gr. ss–j, repeat if necessary.

---

* Not officinal.

† The metal itself is not officinal. Antidote, tannic acid.

Pulvis Antimonialis (antimony oxide, 33
per cent. ; precipitated calcium phos-
phate, 67 per cent.), *James' powder*, . gr. iij–x.
Antimonii Sulphidum Purificatum.
Antimonum Sulphuratum, ... gr. j–v.
Pilulæ Antimonii Compositæ (*Plum-
mer's pills*), aa gr. ⅔ of antimony
and calomel.
Antimonium Oxidum, ... gr. j–ij.
Vinum Antimonii, four grams of tartar
emetic to 1000 c. c. of solution.
Diaphoretic and expectorant dose, . ♏v–xx.
Emetic dose, ... f℥j–iv.
Syrupus Scillæ Compositus contains of ·
tartar emetic 2 gm. to 1000 c. c., . . ♏v–f℥j.

Officinal Name, VERATRUM VIRIDE. *English Name*,
GREEN OR AMERICAN HELLEBORE.

Definition.—*Rhizome* and *rootlets* of Veratrum
viride.†

Natural Order.—Liliaceæ. *Habitat.*—Swamps of
Southern United States.

Alkaloids.—Jervine* and veratroidine.*

Officinal Preparations.

Extractum Veratri Viridis Fluidum, . . ♏j–iv.
Tinctura Veratri Viridis, ... ♏ij–vj.

---

* Not officinal.

† *Norwood's tincture* is made from the green root, and while
not officinal is more powerful than the officinal Tincture of
Veratrum Viride.

Officinal Name, VERATRINA. *English Name*,
VERATRINE.

Definition.—A mixture of alkaloids obtained
from the *seed* of Asagræa officinalis.

Natural Order.—Liliaceæ. *Habitat.*—Mexico.
Used externally for rheumatic pains.

Officinal Preparations.

Oleatum Veratrinæ, two per cent.
Unguentum Veratrinæ, four per cent.

Officinal Name,   *English Name,*
ARNICÆ FLORES.   ARNICA FLOWERS.
ARNICÆ RADIX.   ARNICA ROOT.

Definition.—The *flower-heads, rhizome* and *root-lets* of Arnica montana (leopard's bane).

Natural Order.—Compositæ. *Habitat.*—United
States. ʊ Europe.

Officinal Preparations.

Flowers :
  Tinctura Arnicæ Florum (use externally).   20 %

Root :
  Extractum Arnicæ Radicis, ... gr. v-x.
  Extractum Arnicæ Radicis Fluidum, . ♏x-xx.
  Tinctura Arnicæ Radicis, ... f ℨ ss–ij.
  Emplastrum Arnicæ, 33 per cent. extract.

Officinal Name, ACONITUM. *English Name*, ACONITE.

Definition.—The *tuber* of Aconitum napellus,
(monkshood or wolf's bane).

Natural Order. — Ranunculaceæ. *Habitat.*—
Europe and Asia.

The alkaloid, **Aconitine,*** is almost too strong for internal use.

Dose, gr. ~~$\frac{1}{200}$ $\frac{1}{500}$~~. Used mostly in liniments or ointments.

Officinal Preparations.

Extractum Aconiti, ... gr. ⅙–ss.
Extractum Aconiti Fluidum, ... ♏ss–ij.
Tinctura Aconiti, ... ♏j–v.

Officinal Name, ACIDUM HYDROCYANICUM DILUTUM.
English Name, DILUTED HYDROCYANIC ACID.

Definition.—Two per cent. solution in water of absolute hydrocyanic acid.   Odor and taste those of peach kernels or bitter almonds.

Dose, ♏j–iij.

VEGETABLE ACIDS.

Officinal Name, ACIDUM TARTARICUM.   *English Name,*
TARTARIC ACID.

Definition.—An organic acid usually prepared from argols (sediment of wine); colorless, translucent crystals, sour in taste.

Used only as an ingredient in Seidlitz powder, pulvis effervescens compositus, United States Pharmacopeia.

Officinal Name, ACIDUM CITRICUM.   *English Name,*
CITRIC ACID.

Definition.—An organic acid prepared from

---

* Not officinal.

lemon- or lime-juice. Colorless, translucent, right-rhombic prisms ; odorless ; agreeable, acid taste. Soluble in $\frac{8}{10}$ of its weight of cold water, and in $\frac{1}{20}$ of its weight of boiling water. ·

Officinal Preparation.
Syrupus Acidi Citrici (8 : 1000), ... f3j–iv.

Officinal Name, ACIDUM ACETICUM. *English Name,* ACETIC ACID.

Definition.—Colorless liquid, composed of 36 per cent., by weight, of absolute acetic acid, and 64 per cent. of water.

Used externally as a mild caustic.

Officinal Name, ACIDUM ACETICUM GLACIALE. *English Name,* GLACIAL ACETIC ACID.

Definition.—Nearly or quite absolute acetic acid. Never used internally.

Officinal Name, ACIDUM ACETICUM DILUTUM. *English Name,* DILUTED ACETIC ACID.

Definition.—Six per cent., by weight, of absolute acetic acid ; water to make 100.

Dose, f3j–ij, well diluted.

Used as menstruum for the officinal aceti.

Officinal Name, ACIDUM OXALICUM. *English Name,* OXALIC ACID.

Found naturally in sorrels and in other vegetable life.

Of interest only as a poison.

Antidote.—Lime, calcium carbonate.

Antidote for all the above acids, except oxalic acid, is a mild alkali, followed by oil to prevent, if possible, further corrosion of the intestinal tract. Care must be taken in all cases not to give so much of the antidote that it may itself become poisonous in turn,—hence ammonia, unless very dilute, is contraindicated.

ORDER III.—NUTRIENTS.

FAMILY I.—ASTRINGENTS.

Definition.—Astringents are drugs which cause contractions of those tissues with which they are brought in contact, either directly or through the circulation.

Astringents are (1) vegetable and (2) mineral.

VEGETABLE ASTRINGENTS.

The vegetable astringents all depend for their action on tannic acid.

Officinal·Name, ACIDUM TANNICUM. *English Name*, TANNIC ACID.

Definition.—An organic acid obtained from nut-galls. It is found widely distributed throughout the vegetable kingdom under two forms : (1) *gallo-tannic* (the officinal form); (2) *kino-tannic*. The gallo-tannic strikes a *blue-black* color with the salts of

iron, while kino-tannic gives a *greenish-black* pre-cipitate.

It is a light, yellowish, amorphous powder, occur-ring in scales or masses; possessing a characteristic odor, and strong astringent taste. It coagulates albumen and is the chemic antidote for all the *alka-loids;* forming tannates which are very slow of solu-tion by the intestinal tract. It forms an insoluble tannate of antimony, hence it is also the antidote for antimony in overdose.

Dose, as astringent, gr. iij–v; as hemostatic, gr. x–xx.

Officinal Preparations.

Collodium Stypticum, ... 20 per cent.
Glyceritum Acidi Tannici, ... 20 per cent.
Trochisci Acidi Tannici, ... 1 = gr. j.
Unguentum Acidi Tannici, ... 20 per cent.

Officinal Name, ACIDUM GALLICUM. *English Name*, GALLIC ACID.

Definition.— An organic acid ,prepared from tannic acid by adding one molecule of water of crystallization. Does not coagulate albumen; oc-curs in whitish, silky needles or triclinic prisms; odor-less, astringent taste. Does not precipitate alkaloids as does tannic acid.

Officinal Name, GALLA. *English Name*, NUTGALL.

Definition.—An excrescence on Quercus lusi-tanica (*natural order*, Cupuliferæ), caused by punctures and deposition of ova of Cynips gallæ tinctoriæ (*class*,

13

Insecta ; *order*, Hymenoptera). The best galls come from the Levant, and are usually the size of hickory-nuts.

Rarely used internally.

Officinal Preparations.

Tinctura Gallæ, ... f℥ ss–ij.
Unguentum Gallæ, ... ten per cent.

Officinal Name, CATECHU. *English Name*, CATECHU.

Definition.—An *extract* from the wood of a tree —the Acacia catechu.

Natural Order.—Leguminosæ. *Habitat.*— East India.

Contains tannic acid.

Officinal Preparations.

Tinctura Catechu Composita, ... f℥ ss–ij.
Trochisci Catechu, ... ãã gr. j.

Officinal Name, KINO. *English Name*, KINO.

Definition.—The *inspissated juice* of Pterocarpus marsupium.

Natural Order.—Leguminosæ. *Habitat.*—East Indies.

Occurs in small, dark, brownish-red, shiny pieces ; colors the saliva deep red. Contains kino-tannic acid.

Dose, gr. x–xx.

Officinal Preparation.

Tinctura Kino, ... f℥ ss–ij.

Officinal Name, HÆMATOXYLON.   *English Name*, LOG-
WOOD.

Definition.—The *heart-wood* of Hæmatoxylon
campechianum.

Natural Order.—Leguminosæ.   *Habitat.*—Central America.

Contains hematin and tannic acid.

Gives a blue color in alkaline solution—red in acid.

Officinal Preparation.

Extractum Hæmatoxyli, ... gr. x–xxx.

Officinal Name, KRAMERIA.   *English Name*, RHATANY.

Definition.—The *root* of Krameria triandra and
Krameria ixina.

Natural Order.—Polygaleæ.   *Habitat.* — South
America.

Officinal Preparations.

Extractum Krameriæ, ... gr. v–x.
Extractum Krameriæ Fluidum, . . . ♏v–xxx.
Tinctura Krameriæ, ... f ℥ ss–ij.

Officinal Name, QUERCUS ALBA.   *English Name*, WHITE
OAK.

Definition.—The *inner bark* of Quercus alba.

Natural Order.—Cupuliferæ.   *Habitat.*—United
States.

Used only for astringent washes and lotions.

Officinal Name, ROSA CENTIFOLIA.   *English Name*,
PALE ROSE.

Definition.—The *petals* of Rosa centifolia.

Natural Order.—Rosaceæ.
Used only for its odor.

Officinal Name, ROSA GALLICA. *English Name*, RED
ROSE.

Definition.—The *petals* of Rosa gallica collected
before expanding.
Natural Order.—Rosaceæ.

Officinal Preparations.

Confectio Rosæ, ... ⎫
Extractum Rosæ Fluidum, ... ⎪
Confectio Rosæ, ... ⎬ vehicles.
Mel Rosæ, ... ⎪
Syrupus Rosæ, ... ⎭
Pilulæ Aloes et Mastiches.

Oil of rose is obtained from the *fresh flowers* of
Rosa damascena (*natural order*, Rosaceæ).
This is used in the preparation of **Aqua Rosæ**
and **Unguentum Aquæ Rosæ** (cold cream).

Officinal Name, GERANIUM. *English Name*, CRANES-
BILL, SPOTTED GERANIUM, ETC.

Definition.—The *rhizome* of Geranium macu-
latum.
Natural Order.—Geraniaceæ. *Habitat.*—United
States. Contains large amount of tannic acid.

Officinal Preparation.
Extractum Geranii Fluidum, ... f ʒ ss–j.

Officinal Name, RHUS GLABRA. *English Name,* RHUS GLABRA, SUMAC.

Definition.—The *fruit* of Rhus glabra.

Natural Order.—Anacardieæ. *Habitat.*—United States.

Contains tannic and malic acid.

Officinal Preparation.

Extractum Rhois Glabræ Fluidum (used as a gargle diluted with water).

Mineral Astringents.

Officinal Name, ALUMEN. *English Name,* ALUM.

Definition.—The double sulphate of aluminium and potassium. Soluble in nine parts of water at 59°, and in $\frac{8}{10}$ of a part of boiling water. Gives an acid reaction with litmus.

Dose, astringent, gr. x–xx; emetic, ℥j–iv; in lead poisoning, gr. xx–xl. Also used locally in solution as astringent and styptic.

Officinal Preparation.

Alumen Exsiccatum, sometimes called burnt or dried alum (used externally exclusively).

PLUMBUM. LEAD.*

Found naturally as galena—lead ore, lead sulphide.

---

* Not officinal.

Officinal Preparations.

For internal use :

Plumbi Acetas (sugar of lead), ... gr. ss–v. ✓

For external use :

Liquor Plumbi Subacetatis (*Goulard's* ✓
extract).

Liquor Plumbi Subacetatis Dilutus (lead
water), three per cent. of Goulard's
extract.

Ceratum Plumbi Subacetatis, 20 per cent.
(Goulard's cerate).

Plumbi Carbonas (white lead).

Plumbi Iodidum.

Plumbi Nitras.

Plumbi Oxidum (red lead, litharge).

Unguentum Plumbi Carbonatis, . . . ten per cent.

Unguentum Plumbi Iodidi, ... ten per cent.

Emplastrum Plumbi.

Unguentum Diachylon (lead plaster,
500; olive oil, 490; oil of lavender
flowers, 10).

Poisoning may occur from any of the lead preparations.

Antidote.—Soluble sulphate or dilute sulphuric
acid; alum in chronic poisoning; dilute sulphuric
acid as drink, and iodide of potassium.

BISMUTHUM. BISMUTH.*

Metallic element.

---

* Not officinal.

Officinal Preparations.

Bismuthi Citras, ... gr. j–iij.
Bismuthi et Ammonii Citras, ... gr. j–v.
Bismuthi Subnitras and Bismuthi Sub-
carbonas, gr. v–xv in affections of
stomach, and gr. xv–ʒj in affections
of intestines.

Officinal Name, CERII OXALAS. *English Name*,
CERIUM OXALATE.

Definition.—A white, granular powder; odorless; tasteless. Insoluble in water, alcohol, or ether,
but soluble in sulphuric acid. Used principally to
relieve the vomiting of pregnancy.

Dose, gr. j–iij, in pill t. i. d.

ZINCUM. ZINC.

A metal obtained in the form of carbonates and
sulphides. Officinal in the form of thin sheets,
pencils or fine powder.

Officinal Preparations.

Zinci Oxidum, ... gr. j–v.
Unguentum Zinci Oxidi, ... 20 per cent.
Zinci Acetas (used in solution—gr.
j–v–xx to fʒ—as eye-wash or injec-
tion in gonorrhea).
Zinci Acetas, ... gr. j–v.
Zinci Bromidum, ... gr. j–v.
Zinci Carbonas Præcipitatus, ... external use.
Zinci Chloridum, ... external use.
Liquor Zinci Chloridi, ... external use.
Zinci Iodidum, ... gr. j–v.

Zinci Phosphidum, ... gr. j–v.
Zinci Sulphas (*White Vitriol*). Dose,
astringent, gr. j–ij ; emetic, gr. xxx,
repeated if required.

Antidote. — Alkalies and alkaline carbonates.
Albumen to be used as demulcent.

CUPRUM. COPPER.*

Officinal Preparations.

Cupri Sulphas (*Blue Vitriol*). Dose,
astringent, gr. ¼–j ; emetic, ... gr. iij–v.
Cupri Acetas,* . . . . gr. ⅛–¼.

Copper sulphate is the antidote for poison by
phosphorus.

Antidote.—Potassium ferrocyanide, albumens as
demulcents.

ARGENTUM. SILVER.*

Officinal Preparations.

Argenti Cyanidum (pharmacy).
Argenti Iodidum, ... gr. ss–j in pill.
Argenti Nitras, ... gr. ¼–ss in pill.
Solutions of different strengths are used
as eye-washes, injections, etc.
Argenti Nitras Dilutus (one-half each,
silver and potassium nitrate).
Argenti Nitras Fusus (*Lunar Caustic*,
external use).
Argenti Oxidum, ... gr. ss–j in pill.

---

* Not officinal.

Owing to the readiness with which silver decomposes it should invariably be given in pill, freshly prepared, when intended for internal administration.

Antidote.—Sodium chloride (common salt) forms an insoluble chloride.

FAMILY II.—TONICS.

Tonics are remedies employed to increase the strength and vigor of the body or its organs when depressed from disease or other causes.

MINERAL TONICS.

Officinal Name, FERRUM. *English Name,* IRON.*

Officinal Preparations.

Ferri Carbonas Saccharatus, ... gr. v–ʒ ss.
Ferri Chloridum.
Ferri Citras.
Ferri et Ammonii Citras, ... gr. ij–v.
Ferri et Ammonii Sulphas, ... gr. ij–v.
Ferri et Ammonii Tartras, ... gr. j–x.
Ferri et Potassii Tartras, ... gr. v–x.
Ferri et Quininæ Citras, ... gr. v–x.
Ferri et Quininæ Citras Solubilis.
Ferri et Strychninæ Citras, one per
　cent., ... gr. j–v.
Ferri Hypophosphis, ... gr. v–x.
Ferri Iodidum Saccharatum, ... gr. v–xv.
Ferri Lactas, ... gr. v–x.
Ferri Oxidum Hydratum.

---

* Officinal in the form of fine, bright, and non-elastic wire.

14

**Ferri Oxidum Hydratum cum Mag-
nesia** (used in *arsenical poisoning*).

Ferri Phosphas Solubilis, ... gr. ij–v.

Ferri Pyrophosphas Solubilis, ... gr. ij–v.

Ferri Sulphas (*green vitriol*), . . . ⎫

Ferri Sulphas Exsiccatus, ... ⎬ gr. ss–ij.

Ferri Sulphas Granulatus, ... ⎭

Ferri Valerianas, ... gr. j–iij.

Besides the above, the most commonly used preparations of iron are :

Ferrum Reductum (reduced iron, *Que-
venne's iron*), ... gr. ij–v in pill.

Massa Ferri Carbonatis (*Vallet's mass*), gr. iij–x in pill.

Liquor Ferri Tersulphatis (used with
Ferri Oxidum Hydratum and Ferri
Oxidum Hydratum cum Magnesia, as
antidotes in arsenical poisoning).

Liquor Ferri Subsulphatis (*Monsel's
solution*) used locally as styptic.

Tinctura Ferri Chloridi, ... ℥v–xxx.

Syrupus Ferri Iodidi, ... ℥v–xxx.

Liquor* Ferri et Ammonii Acetatis
(*Basham's mixture*, tonic and diuretic), ... f ʒj–iv.

Ferrum Dialysatum,† . . . . . . f ʒss–f ʒj.

As antidote for arsenic, ... f ℥ss.

Liquor Ferri Acetatis, ... external use.

Liquor Ferri Chloridi, ... external use.

Liquor Ferri Citratis, ... ℥v–xv.

Liquor Ferri Nitratis, ... ℥ij–x.

Syrupus Ferri Iodidi, ... f ʒj–ij.

---

* Mistura (Pharm., 1880). † Not officinal.

Syrupus Ferri, Quininæ, et Strychninæ
Phosphatum, ... f℥j–ij.
Vinum Ferri Amarum, ... f℥j–ij.
Vinum Ferri Citratis, ... f℥j–ij.
Pilulæ Aloes et Ferri, ... j pill.
Pilulæ Ferri Carbonatis, ... j–iij pills.
Pilulæ Ferri Iodidi, ... j–iij pills.
Trochisci Ferri (1 = gr. v of ferric hy-
drate), ... j–v troches.
Emplastrum Ferri.

MANGANUM. MANGANESE.*

Officinal Preparations.
Mangani Oxidum Nigrum (1880), . . gr. j–x in pill.
Mangani Sulphas, ... gr. j–v.
Mangani Dioxidum (same as Mangani
Oxidum Nigrum, of Pharm., 1880)..

Officinal Name, ACIDUM SULPHURICUM (H_2SO_4).
English Name, SULPHURIC ACID (called sometimes OIL OF VITRIOL).

Definition.—A liquid composed of not less than
92.5 per cent., by weight, of absolute sulphuric acid,
and not more than 7.5 per cent. of water.

Used internally only if diluted. Poisoning causes
charring and blackening of the tissues.

Officinal Preparations.
Acidum Sulphuricum Dilutum (ten per
cent. of officinal acid), ... ♏v–xx further di-
Acid Sulphuricum Aromaticum (*elixir*   . [luted.
of vitriol), ten per cent., ... ♏v–xx.

---

* Not officinal.

Officinal Name, ACIDUM HYDROCHLORICUM (HCl).
English Name, HYDROCHLORIC OR MURIATIC
ACID.

Definition.—A colorless liquid, composed of 32 per cent., by weight, of absolute hydrochloric acid.

Not used internally. Poisoning leaves a yellow stain on the tissues.

Officinal Preparation.

Acidum Hydrochloricum Dilutum (ten
per cent. of absolute acid), . . . ℥x–xx.

Officinal Name, ACIDUM NITRICUM (HNO$_3$). *English Name,* NITRIC ACID.

Definition.—A liquid composed of 68 per cent., by weight, of absolute nitric acid, and 32 per cent. of H$_2$O.

Not used internally. Poisoning causes deep orange-yellow staining of the tissues.

Officinal Preparation.

Acidum Nitricum Dilutum (ten per cent.
of absolute acid), ... ℥x–xx.

Officinal Name, ACIDUM NITROHYDROCHLORICUM.
English Name, NITROHYDROCHLORIC ACID.

Definition.—Consists of 18 parts nitric acid and 82 parts hydrochloric acid.

Officinal Preparation.

Acidum Nitrohydrochloricum Dilutum
(nitric acid, four parts ; hydrochloric
acid, 18 parts ; and water, 78 parts), . ℥x–xx.

208   FOR THERAPEUTIC NOTES.

Officinal Name, ACIDUM LACTICUM. *English Name,* LACTIC ACID.

Definition.—The acid of milk-sugar, composed of 75 per cent. of absolute lactic acid.

Dose, ♏v–xxx.

Officinal Name, PHOSPHORUS. *English Name,* PHOS-PHORUS.

Definition.—Translucent, almost colorless, solid; waxy in appearance and consistency. Has an odor resembling garlic, and characteristic taste (should never be tasted unless greatly diluted). Takes fire spontaneously on exposure to the air.

Dose, gr. $\frac{1}{100}$.

Officinal Preparations.

Oleum Phosphoratum (one per cent. of phosphorus in oil of almonds and ether), ... ♏j–iij.

Pilulæ Phosphori, ... āā gr. $\frac{1}{100}$.

Spiritus Phosphori (used to make elixir).

Zinci Phosphidum, ... gr. $\frac{1}{30}$–¼.

Elixir Phosphori (two per cent. of spirit), ♏x.

Antidotes.—Sulphate of copper, French oil of turpentine.

FAMILY III.—ALTERATIVES.

Medicines which in some way seem to alter the nutrition and increase the strength and health in various pathologic states.

ARSENUM. ARSENIC.*

Definition.—A metallic element found often with other metals as an arsenide. Black arsenic, sometimes called cobalt.

Officinal Preparations.

Acidum Arsenosum (white arsenic), . gr. 1/30.
Liquor Acidi Arsenosi, ... ⎫
Liquor Potassii Arsenitis (*Fow-*
  ler's solution), ... ⎬ ♏ij–x.
Liquor Sodii Arsenatis—each of
  the three above liquors contain
  one per cent. of arsenous acid, . ⎭
Sodii Arsenas, ... gr. 1/12–1/4.
Arseni Iodidum, ... gr. 1/8.
Liquor Arseni et Hydrargyri Iodidi
  (*Donovan's solution*, one per cent.
  each of arsenic and mercuric iodide), ♏ij–x.

Antidote.—Freshly prepared hydrated oxide of iron with magnesia in large amount, emetics, etc.

Officinal Name, HYDRARGYRUM. *English Name,*
MERCURY.

Definition.—A liquid element obtained from the sulphide cinnabar. Specific gravity, 13.5.

Officinal Preparations.

For internal use :

· Hydrargyrum cum Creta (gray powder),
  38 per cent. of mercury, small doses as
  laxative, large doses as antisyphilitic, gr. j–v–xx.

V

---

* Not officinal.

For internal use :

Massa Hydrargyri (blue mass), 33 per
cent. mercury, ... gr. j–x.   ✓

Hydrargyri Chloridum Corrosivum (bi-
chloride, corrosive sublimate), . . . gr. $\frac{1}{50}-\frac{1}{10}$.

Hydrargyri Chloridum Mite (calomel), . gr. ss–xx.   ✓

Hydrargyri Cyanidum, ... gr. $\frac{1}{20}-\frac{1}{12}$.

Hydrargyri Iodidum Rubrum (red
iodide), ... gr. $\frac{1}{30}-\frac{1}{16}$.   ✓

Hydrargyri Iodidum Viride (Pharm.,
1880), same as Hydrargyri Iodidum
Flavum, ... gr. $\frac{1}{4}$–j.   ✓

Hydrargyri Iodidum Flavum (prot-
iodide, yellow or green iodide), . . gr. $\frac{1}{4}$–j.

Pilulæ Catharticæ Compositæ (see Colocynth), ... j–iij pills.

Hydrargyri Subsulphas Flavus (*turpeth
mineral*), as emetic, ... gr. ij–v.

Pilulæ Antimonii Compositæ (*Plummer's
pill*, calomel and sulphurated antimony).

Antidote.—Albumen (white of egg), milk, wheat
flour, emetic.

For external use only :

Emplastrum Hydrargyri.

Emplastrum Ammoniaci cum Hydrargyri.

Unguentum Hydrargyri (blue ointment),
45 per cent. mercury.

Unguentum Hydrargyri Ammoniati, ten per cent.

Unguentum Hydrargyri Oxidi Rubri, ten per cent.

Unguentum Hydrargyri Oxidi Flavi, ten per cent.

Unguentum Hydrargyri Nitratis (citrine ointment).

Hydrargyri Ammoniatum (white precipitate).

For external use only :

Hydrargyri Oxidum Rubrum (red precipitate).

Hydrargyri Oxidum Flavum.

Oleatum Hydrargyri (yellow oxide), ten per cent.

Liquor Hydrargyri Nitratis (caustic).

Hydrargyri Subsulphidum Rubrum (cinnabar), used in fumigation.*

Officinal Name, AURI ET SODII CHLORIDUM. *English Name*, CHLORIDE OF GOLD AND SODIUM.

Definition.—A mixture of equal parts, by weight, of dry gold chloride and sodium chloride.

Dose, gr. $\frac{1}{12}$–$\frac{1}{4}$ in pill.

Officinal Name, IODUM. *English Name*, IODINE.

Definition.—Non-metallic element made from the ashes of seaweed ; forms blue color with starch, which is the antidote in poisoning by it.

Officinal Preparations.

Iodi, ...	gr. ¼–j.
Liquor Iodi Compositus (*Lugol's solution*), iodine, 5 ; potassium iodide, 10 ; water, 85, ...	♏ iij–x.
Ammonii Iodidum, ...	gr. ij–x.
Potassii Iodidum, ...	gr. v–Ʒ ss.
Strontii Iodidum, ...	gr. v–Ʒ ss.
Syrupus Acidi Hydriodici, ...	fƷ ss–ij.

Externally only :

Tincturæ Iodi, ...	eight per cent.
Unguentum Iodi, ...	four per cent.
Unguentum Potassii Iodidi, ...	twelve per cent.

---

* Not officinal.

Officinal Name, IODOFORMUM. *English Name*,
IODOFORM.

Definition.—Small yellow crystals, characteristic odor, and iodine-like taste ; slightly soluble in water; soluble in alcohol, chloroform, and ether ; volatile.

Used chiefly as an antiseptic.

Dose, gr. j–iv.

Officinal Preparation.

Unguentum Iodoformi, ... ten per cent.

Aristol and **iodol** * may be used in the same way as iodoform.

Officinal Name, OLEUM·MORRHUÆ, OLEUM JECORIS
ASELLI. *English Name*, COD-LIVER OIL.

Definition.—A fixed oil obtained from the fresh livers of Gadus morrhua and other species of Gadus.

Contains gadium, iodine, chlorine, bromine, and fatty acids.

Dose, f3j–f3ss, t. i. d. ; mostly given in emulsion.

Officinal Name, ACIDUM PHOSPHORICUM. *English
Name*, PHOSPHORIC ACID.

Definition.—A liquid, containing not less than 85 per cent., by weight, of absolute orthophosphoric acid. For external use and pharmacy only.

Officinal Preparation.

Acidum Phosphoricum Dilutum (ten per
cent. of absolute acid), ... ♏v–xxx.

---

* Not officinal.

218   FOR THERAPEUTIC NOTES.

Officinal Name,	*English Name,*
COLCHICI SEMEN.	COLCHICUM SEED.
COLCHICI RADIX.	COLCHICUM ROOT.

Definition.—The *seeds* and *corm* of Colchicum autumnale, or meadow saffron.

Natural Order.—Liliaceæ.   *Habitat.*—Europe.

Contains the alkaloid colchicine, which is its active principle.

Used in the form of the salicylate.

Dose, gr. $\frac{1}{60}$.

Seed :   *Officinal Preparations.*

    Extractum Colchici Seminis Fluidum, . ♏ij–vi.

    Tinctura Colchici Seminis, ... f℥ss–j.

    Vinum Colchici Seminis, ... ♏v–xxx.

Root :

    Extractum Colchici Radicis, ...   gr. ss–ij.

    Extractum Colchici Radicis Fluidum, . ♏ij–v.

    Vinum Colchici Radicis, ... ♏v–xv.

Officinal Name, SARSAPARILLA.   *English Name,* SARSA-
PARILLA.

Definition.—The *root* of Smilax officinalis, Smilax medica, and other varieties of Smilax.

Natural Order. — Liliaceæ.   *Habitat.* — Mexico, South and Central America.

Contains the glucoside *smilacin.*

Officinal Preparations.

Decoctum Sarsaparillæ Compositum, . f℥ij–iv.

Extractum Sarsaparillæ Fluidum, . . . f℥j.

Extractum Sarsaparillæ Fluidum Com-

    positum, ... f℥j.

Syrupus Sarsaparillæ Compositus,   . . f℥ss.

Officinal Name,	*English Name,*
GUAIACI LIGNUM.	GUAIACUM WOOD.
GUAIACI RESINA.	GUAIAC.

Definition.—The *heart-wood* of Guajacum offici-nale and Guaiacum sanctum, an evergreen tree of South America. The resin is obtained from the wood, and the preparations are made from it.

Natural Order.—Zygophylleæ.

Dose, gr. v–xx.

Officinal Preparations.

Tinctura Guaiaci, ... f ℥ ss–ij.
Tinctura Guaiaci Ammoniata, ... f ℥ ss–ij.

Officinal Name, MEZEREUM.   *English Name,*
MEZEREUM.

Definition.—*Bark* of Daphne mezereum, and other species of Daphne.

Natural Order.—Thymelæaceæ.   *Habitat.*—Europe.

Contains the neutral, bitter glucoside *daphnin.*

Officinal Preparation.

Extractum Mezerei Fluidum (used only for pharmaceutic purposes) is used in both the compound decoction and compound fluid extract of sarsaparilla.

Officinal Name, SASSAFRAS.   *English Name,*
SASSAFRAS.

Definition.—The *bark* of the *root* of Sassafras variifolium.

Natural Order.—Laurineæ.   *Habitat.*—Europe and United States.

Contains a volatile oil, used mostly for flavoring.

Officinal Name, TARAXACUM. *English Name*,
DANDELION.

Definition.—The *root* of Taraxacum officinale.

Natural Order.—Compositæ. *Habitat.*—Indigenous.

Contains *taraxacin*, a bitter principle.

Officinal Preparations.

Extractum Taraxaci, ... gr. xx–ʒj.
Extractum Taraxaci Fluidum, ... fʒj–ij.

ICHTHYOL.*

Definition.—A substance obtained by the distillation of a bituminous, sulphurous mineral deposit, found in North Germany, due to decomposition of fossil fish. Is now made synthetically. Contains about ten per cent. of sulphur.

Used externally as ointment; internally, gr. j–iij in pill or capsule.

FAMILY IV.—ANTIPERIODICS.

These remedies—which have for their type quinine—are used to overcome periodic fevers ; *e. g.*, malaria.

Officinal Name, CINCHONA.

Definition.—The *bark* of any species of Cinchona which contains five per cent. of total alkaloids, and at least 2.5 per cent. of quinine ($C_{20}H_{24}N_2O_2 + H_2O$).

Natural Order.—Rubiaceæ. *Habitat.*—Peru and Bolivia.

---

* Not officinal.

The bark of **Cinchona calisaya, Cinchona officinalis,**—*English name*, yellow cinchona,—is used in preparing the following:

Officinal Preparations.

Extractum Cinchonæ, ... gr. v–xx.
Extractum Cinchonæ Fluidum, . . . f ʒ j.
Tinctura Cinchonæ, ... f ʒ j–iv.
Infusum Cinchonæ, ... f ʒ j–ij.

Officinal Name, CINCHONA RUBRA. *English Name*, RED CINCHONA.

Definition.—The *bark* of Cinchona succirubra. *Natural Order.*—Rubiaceæ.

Officinal Preparation.

Tinctura Cinchonæ Composita (Huxham's tincture), ... f ʒ j–iv.

The alkaloids of cinchona in the order of their potency are (1) quinine, (2) cinchonine, (3) quinidine, (4) cinchonidine.

Officinal Alkaloids and Salts.

Quinina, ... }	As a tonic, gr. j–
Quininæ Sulphas, ...	iij ; as an anti-
Quininæ Bisulphas (more soluble than	pyretic, gr. v–
the sulphate), ...	xx ; as an anti-
Quininæ Hydrochloras, ...	periodic, gr. v–
Quininæ Hydrobromas, ...	xxx.
Quininæ Valerianas, ...	gr. j–v.
Cinchonina, ... }	Doses about one-
Cinchoninæ Sulphas, ... }	third larger than
Cinchonidinæ Sulphas, ... }	quinine.

Quinine and quinidine, or any salt of either, if treated with fresh chlorine or bromine water form an *emerald-green* precipitate if ammonia water is added to the solution.   Cinchonine and cinchonidine, or their salts, form a *white* precipitate when thus treated.

Aqueous solutions of quinine, quinidine, and their salts, when acidulated with sulphuric acid, produce a pale *bluish* efflorescence.   A weak solution of cinchonine or its salts should not exhibit more than a pale *yellow* color.   Morphine imparts to sulphuric acid only a pale *yellow* tinge.   Quinine and quinidine, and their salts, should not cause a *red* color with nitric acid as does morphine.

WARBURG'S TINCTURE.*

Definition.—A dark-brown liquid containing numerous ingredients much used in the pernicious malarial fevers of India. (See formula in United States Dispensatory.)

Dose, fℨss.

Officinal Name, EUCALYPTUS.   *English Name*, EUCALYPTUS.

Definition.—The *leaves* of Eucalyptus globulus collected from the older parts of the tree.

Natural Order.—Myrtaceæ.   *Habitat.*—Australia.

Officinal Preparations.

Extractum Eucalypti Fluidum, ... fℨj–ij.
Oleum Eucalypti, ... ♏v–xx.

---

* Not official.

ACIDUM PICRICUM. PICRIC ACID.*

Trinitrophenol, $C_6H_2(NO_2)_3OH$.

Definition.—Made by dissolving crystals of carbolic acid in strong sulphuric acid, and adding nitric acid to the solution.

Used mostly in the arts.

FAMILY V.—ANTIPYRETICS.

Remedies which do not affect the normal temperature but cause a reduction of the temperature in febrile conditions.

Officinal Name, ACIDUM CARBOLICUM. *English Name*, CARBOLIC ACID.

Definition.—A constituent of coal-tar, obtained by distillation and then purified ; also called *phenic* and *phenylic acid.*

The crude form is used as a disinfectant. Occurs in the form of needle-shaped crystals, white when pure but inclined to turn red.

Used internally in **doses** of gr. j.

Officinal Preparation.

Unguentum Acidi Carbolici, ten per cent.

Antidote.—Soluble sulphate (magnesium sulphate) rapidly administered in large amount.

---

* Not officinal.

Officinal Name, CREOSOTUM.   *English Name*,
CREOSOTE.

Definition.—A mixture of phenols, chiefly of guaiacol (the value of preparation depends on the amount of guaiacol it contains), and creosol, obtained by the distillation of wood-tar.   That from beech-wood (Fagus sylvatica, *natural order*, Cupuliferæ) is preferred.

Dose, ♏ j-iij.

Officinal Preparation.

Aqua Creosoti, one per cent., ... f ℥ j-iv.

GUAIACOL.*

Definition.—A syrupy liquid found in creosote ; often used in the form of the carbonate.

Dose, gr. v.

Guaiacol itself is used in doses of five drops— ♏ xx-xl daily.

Officinal Name, MENTHOL.   *English Name*, MENTHOL.

Definition.—A stearopten obtained from the officinal oil of peppermint (Mentha piperita) or other menthæ.

Used as a local anesthetic by rubbing it on the part to be anesthetized.

Officinal Name, THYMOL.

Definition.—A phenol obtained from the oil of thyme.   It is a local anesthetic and antiseptic, and often used as a spray in throat and mouth affections.

---

* Not officinal.

Officinal Name, RESORCINUM.   *English Name*,
RESORCIN.

Definition.—A diatomic phenol, antiseptic and antifermentative.

Dose, gr. ij–v.

Officinal Name, ACIDUM SALICYLICUM.   *English Name*,
SALICYLIC ACID.

Definition.—An organic acid found in most plants, but generally prepared synthetically from carbolic acid.   Soluble in 450 parts of water and in 2 4 of alcohol.

Dose, gr. x–ʒj.

Officinal Preparations.

Sodii Salicylas, ... gr. v–ʒ ss.
Lithii Salicylas, ... gr. j–x.
(ʒj–jss of the Sodii Salicylas in twenty-four hours, preferably administered in milk.)

Officinal Name, OLEUM GAULTHERIÆ.   *English Name*,
OIL OF WINTERGREEN.

Definition.—A volatile oil distilled from the *leaves* of the Gaultheria procumbens.

Natural Order.—Ericaceæ.   *Habitat.*—United States.

Consists almost entirely of *methyl salicylate*, to which its virtue is due.

Dose, ℥x–xv, in capsule or emulsion.

Officinal Preparation.

Spiritus Gaultheriæ, ... fʒ ss–j.
16

Officinal Name, SALICINUM. *English Name*, SALICIN.

Definition.—A neutral principle obtained from several species of Salix (willow). Consists of white, silky, crystalline needles; odorless, but of very bitter taste. Soluble in hot or cold water.

Dose, gr. x–ʒj.

Officinal Name, BENZOINUM. *English Name*, BENZOIN.

Definition.—A balsamic resin obtained from Styrax benzoin.

Natural Order.—Styraceæ. *Habitat.*—Peru.
Contains *benzoic acid, a volatile oil and a resin.*

Officinal Preparations.

Adeps Benzoinatus, ... external use.
Tinctura Benzoini, ... fʒj–ij.
Tinctura Benzoini Composita, . . . fʒj–ij.

Officinal Name, SALOL. *English Name*, SALOL.

Definition.—The salicylic ether of phenol. A white, crystalline powder resembling in odor oil of wintergreen; tasteless; almost insoluble in water.

Dose, gr. v–xv, t. i. d.

ANTIPYRIN.*

Definition.—A white, odorless powder, having a slightly bitter taste. Obtained synthetically and by the distillation of coal-tar.

Dose, gr. x–xx.

---

* Not official.

ANTIFEBRIN (ACETANILIDUM,* PHENACETIN). †

Definition.—Coal-tar products, all resembling antipyrin in appearance and effect.

Dose, gr. v–xv, repeated if necessary.

THALLIN. †

Definition.—A synthetically prepared alkaloid ; antipyretic.

Dose, gr. v–x.

FAMILY I.—STOMACHICS.

Substances which, by increasing the activity of the glands of the gastro-intestinal tract, facilitate digestion.

The simple bitters are all of vegetable origin, of bitter taste, and while they stimulate markedly the mucous membrane of the gastro-intestinal tract, have practically no effect on the general system.

They include **quassia** and **calumba,** whose preparations contain no *tannic acid* and may therefore be used with iron preparations.

Almost if not all other vegetable preparations contain tannic acid, and are therefore incompatible with iron preparations.

SIMPLE BITTERS.

Officinal Name, QUASSIA.

Definition.—The *wood* of Picræna excelsa.

---

* Acetanilidum is officinal.　　　　† Not officinal.

Natural Order.—Simarubeæ. *Habitat.*—Jamaica.
Contains *quàssin,* a neutral, bitter principle.

Officinal Preparations.

Extractum Quassiæ, ... gr. j–v.
Extractum Quassiæ Fluidum, ... ℳx–xxx.
Tinctura Quassiæ, ... fℨss–ij.

An infusion is often used as an enema in treatment
for seat-worms.

Officinal Name, GENTIANA. *English Name,* GENTIAN.

Definition.—The *root* of Gentiana lutea (yellow
alpine gentian).
Natural Order.—Gentianeæ. *Habitat.*—Europe.
Contains the active principle *gentiopikrin* and gen-
tisic acid.

Officinal Preparations.

Extractum Gentianæ, ... gr. ij–x.
Extractum Gentianæ Fluidum, ... ℳx–xxx.
Tinctura Gentianæ Composita, . . . fℨj–iv.

Officinal Name, HYDRASTIS. *English Name,* GOLDEN
SEAL.

Definition.—The *rhizome* and *rootlets* of Hydrastis canadensis, tumeric root, etc.
Natural Order.—Ranunculaceæ. *Habitat.*—Indigenous.
Contains the alkaloids **hydrastine** and **berberine.**

Officinal Preparations.

Extractum Hydrastis Fluidum, . . . ℔x–f℥j.
Glyceritum Hydrastis, ... f℥j.
Tinctura Hydrastis, ... f℥ss–ij.
Hydrastininæ Hydrochloras, ... gr. ¼.

Officinal Name, CALUMBA. *English Name*, COLUMBO.

Definition.—The *root* of Jateorhiza palmata.

Natural Order.—Menispermaceæ. *Habitat.*—Africa.

Contains the alkaloid **berberine**, which is found in many plants, and *columbin*, a bitter principle; but *no* tannic acid.

Officinal Preparations.

Extractum Calumbæ Fluidum, ... ℔xv–f℥ss.
Tinctura Calumbæ, ... f℥j–ij.

Officinal Name, EUPATORIUM. *English Name*, THOR-
OUGHWORT, BONESET.

Definition.—The *leaves* and *flowering tops* of Eupatorium perfoliatum.

Natural Order.—Compositæ. *Habitat.*—Indigenous.

Officinal Preparation.

Extractum Eupatorii Fluidum, ... f℥j–iv.

Officinal Name, PRUNUS VIRGINIANA. *English Name*,
WILD CHERRY BARK.

Definition.—The *inner bark* of Prunus serotina, collected in autumn.

Natural Order.—Rosaceæ.   *Habitat.*—United States.

It contains the glucoside *amygdalin*, tannic acid, bitter extractives, and emulsin.

Officinal Preparations.

Extractum Pruni Virginianæ Fluidum, f ʒ ss.
Infusum Pruni Virginianæ, ... f ℥ ij.
Syrupus Pruni Virginianæ, ... vehicle.

Officinal Name, CHIRATA.

Definition.—The entire *plant*, Swertia chirata.
Natural Order.—Gentianeæ. *Habitat.*—India.

Officinal Preparations.

Extractum Chiratæ Fluidum, ... ℳ x–xxx.
Tinctura Chiratæ, ... f ʒ ss–j.

AROMATICS.

Used mostly as carminatives (to expel flatus), to disguise the taste of other medicines, and to prevent the griping of purgatives. They all contain a volatile oil, to which their activity is due.

Officinal Name, CINNAMOMUM ZEYLANICUM. *English Name,* CEYLON CINNAMON (CINNAMOMUM, Pharm., 1880).

Definition.—The *inner bark* of the shoots of Cinnamomum zeylanicum.
Natural Order.—Laurineæ.   *Habitat.*—Ceylon.

Officinal Preparations.

Oleum Cinnamomi, ... ♏︎j–v.
Aqua Cinnamomi, ... vehicle.
Spiritus Cinnamomi, ... ♏︎x–f℥ss.
Tinctura Cinnamomi, ... f℥ss–ij.
Pulvis Aromaticus, contains cinnamon,
    ginger, nutmeg, and cardamom, . . gr. x–xx.
Extractum Aromaticum Fluidum, . . ♏︎x–xx.

Officinal Name, MYRISTICA. *English Name,* NUTMEG.

Definition.—The *seed* of Myristica fragrans, deprived of its testa.

Natural Order.—Myristicaceæ. *Habitat.*—West Indies and various islands of Asia.

Used in pulvis aromaticus and tinctura lavandulæ composita.

Officinal Preparations.

Oleum Myristicæ, ... ♏︎ij–v.
Spiritus Myristicæ, ... f℥ss–ij.

Officinal Name, CARYOPHYLLUS. *English Name,*
CLOVES.

Definition.—The unexpanded *flowers* of Eugenia aromatica.

Natural Order.—Myrtaceæ. *Habitat.*—Molucca Islands, India.

Used in tinctura lavandulæ composita.

Officinal Preparation.

Oleum Caryophylli, ... ♏︎j–v.

Officinal Name, PIMENTA. *English Name*, ALLSPICE.

Definition.—The nearly *ripe fruit* of Pimenta officinalis.

Natural Order.—Myrtaceæ. *Habitat.*—West Indies.

Officinal Preparation.

Oleum Pimentæ, ... ♏j–v.

Officinal Name, CARDAMOMUM. *English Name*, CARDA-
MOM, OR CARDAMOM SEEDS.

Definition.—The *fruit* of Elettaria repens.

Natural Order. — Scitamineæ. *Habitat.* — East Indies.

Used in pulvis aromaticus.

Officinal Preparations.

Tinctura Cardamomi, ... f℥ss–j.

Tinctura Cardamomi Composita, . . f℥ss–j.

Officinal Name, PIPER. *English Name*, BLACK PEPPER.

Definition.—The *unripe fruit* of Piper nigrum.

Natural Order.—Piperaceæ. *Habitat.*—East Indies.

Contains an acrid resin, the alkaloid piperine, and a volatile oil.

Officinal Preparation.

Oleoresina Piperis, ... ♏ss–ij.

Officinal Name, CAPSICUM. *English Name*, RED,
CAYENNE, AFRICAN, PEPPER.

Definition.—The *fruit* of Capsicum fastigiatum.

Natural Order. — Solanaceæ. *Habitat.* — West Indies.

Contains a resin, fixed and volatile oils.

May be used in doses of gr. ij-v, in pill.

Officinal Preparations.

Extractum Capsici Fluidum, ... ℳij–x.
Oleoresina Capsici, ... gr. ss–j.
Tinctura Capsici, ... ℳv–xxx.
Emplastrum Capsici.

Officinal Name, ZINGIBER. *English Name,* GINGER.

Definition.—The *rhizome* of Zingiber officinale.

Natural Order.—Scitamineæ. *Habitat.*—East and West Indies.

Contains a volatile oil and an acrid resin.

Used in pulvis aromaticus and in pulvis rhei compositus.

Officinal Preparations.

Oleoresina Zingiberis, ... ℳss–ij.
Extractum Zingiberis Fluidum, . . . ℳv–xv.
Tinctura Zingiberis, ... ʒss–j.
Syrupus Zingiberis, ... vehicle.
Trochisci Zingiberis, ... j = ℳij of tincture.

Officinal Name, FŒNICULUM. *English Name,* FENNEL.

Definition.—The *fruit* of Fœniculum capillaceum.

Natural Order.—Umbelliferæ. *Habitat.*—South England.

17

Officinal Preparations.

Aqua Fœniculi, ... f℥j-f℥ss.
Oleum Fœniculi, ... ℳj-v.

Officinal Name, OLEUM CAJUPUTI.   *English Name*,
OIL OF CAJUPUT.

Definition.—A volatile oil distilled from the *leaves* of Melaleuca leucadendron.

Natural Order.—Myrtaceæ.   *Habitat.*—Molucca Islands.

Dose, ℳv-xv.

Used as parasiticide, externally.

Officinal Name, CARUM.   *English Name*, CARAWAY.

Definition.—The *fruit* of Carum carvi.

Natural Order. — Umbelliferæ.   *Habitat.* — Europe.

Used in compound tincture of cardamom.

Officinal Preparation.

Oleum Cari, ... ℳj-v.

Officinal Name, CORIANDRUM.   *English Name*,
CORIANDER.

Definition.—The *fruit* of Coriandrum sativum.

Natural Order.—Umbelliferæ.   *Habitat.* — Europe.

Dose, gr. v-xxx.

Officinal Preparation.

Oleum Coriandri, ... ℳj-v.

Officinal Name, ANISUM. *English Name*, ANISE.

Definition.—The *fruit* of Pimpinella anisum.

Natural Order.—Umbelliferæ. *Habitat.* — Europe.

Officinal Preparations.

Oleum Anisi, ... ♏ j–v.
Spiritus Anisi, ... f℥ ss–ij.
Aqua Anisi, ... vehicle.

Officinal Name, OLEUM SASSAFRAS. *English Name*, OIL OF SASSAFRAS.

Definition.—A volatile *oil* distilled from Sassafras.

Dose, ♏ ij–v, mostly as flavoring.

Officinal Name, AURANTII DULCIS CORTEX. *English Name*, SWEET ORANGE PEEL.

Definition.—The *rind* of the *fruit* of Citrus aurantium.

Natural Order. — Rutaceæ. *Habitat.* — Europe and United States.

Officinal Preparations.

Oleum Aurantii Corticis, ...	
Spiritus Aurantii, ...	vehicles and
Spiritus Aurantii Compositus, . . .	flavors.
Tinctura Aurantii Dulcis, ...	
Syrupus Aurantii, ...	

Officinal Name, AURANTII AMARI CORTEX. *English Name*, BITTER ORANGE PEEL.

Definition.—The *rind* of the *fruit* of Citrus vulgaris.

Natural Order.—Rutaceæ.

Officinal Preparations.

Extractum Aurantii Amari Fluidum, . f ʒ j.
Tinctura Aurantii Amari, ... f ʒ j–ij.

Officinal Name, AURANTII FLORES (Pharm., 1880).
English Name, ORANGE FLOWERS.

Definition.—The *flowers* of both the foregoing species.

Officinal Preparations.

Syrupus Aurantii Florum, ... vehicle.
Aqua Aurantii Florum Fortior, ... vehicle.
Oleum Aurantii Florum, ... flavor.

Officinal Name, LAVANDULA (Pharm., 1880). *English Name*, LAVENDER.

Definition.—The *fresh leaves* and *tops* of Lavandula officinalis.

Natural Order.—Labiatæ.

Officinal Preparations.

Oleum Lavandulæ Florum, ... ɱ j–v.
Tinctura Lavandulæ Composita, . . . f ʒ ss–ij.
Spiritus Lavandulæ, ... f ʒ ss–ij.

Officinal Name, SALVIA. *English Name*, SAGE.

Definition.—The *leaves* of Salvia officinalis.

Natural Order. — Labiatæ. *Habitat.* — United States.

Often used as a gargle in shape of an infusion.

Officinal Name, ROSMARINUS (Pharm., 1880). *English Name*, ROSEMARY.

Definition.—The *leaves* of Rosmarinus officinalis.

The oil is sometimes used in two to five drop doses.

Officinal Name, MENTHA PIPERITA. *English Name*, PEPPERMINT.

Definition.—The *leaves* and *tops* of Mentha piperita.

Officinal Name, MENTHA VIRIDIS. *English Name*, SPEARMINT.

Definition.—The *leaves* and *tops* of Mentha viridis.

Both Mentha piperita and Mentha viridis belong to *Natural order*, Labiatæ. *Habitat.*—United States.

The oil, water, and spirit of both peppermint and spearmint are officinal.

Officinal Preparations.

Oleum Menthæ Piperitæ, ... ⎫
Oleum Menthæ Viridis, ... ⎬ vehicles.
Aqua Menthæ Piperitæ, ... ⎪
Aqua Menthæ Viridis, ... ⎭
Spiritus Menthæ Piperitæ, ... f ʒ ss.
Spiritus Menthæ Viridis, ... f ʒ ss.

Officinal Name, MELISSA. *English Name*, BALM.

Definition.—The *leaves* and *tops* of Melissa officinalis.

Natural Order.—Labiatæ.

Used in infusion.

Officinal Name, CALAMUS. *English Name*, SWEET
FLAG.

Definition.—The *rhizome* of Acorus calamus.

Natural Order.—Aroideæ.　*Habitat.*—United
States.

Officinal Preparation.

Extractum Calami Fluidum, ... f ℥ ss–j.

AROMATIC BITTERS.

Officinal Name, ANTHEMIS. *English Name*, CHAMO-
MILE, OR CHAMOMILE FLOWERS.

Definition.—The *flower-heads* of Anthemis nobilis.

Natural Order.—Compositæ.　*Habitat.*—Europe
and America.

Mostly used as an infusion, the so-called chamomile tea (German Brnst-Thee), breast-tea.

Dose, ℥ss of the flowers to Oj of water.

Officinal Name, SERPENTARIA. *English Name*, VIR-
GINIA SNAKEROOT.

Definition.—The *rhizome* and *rootlets* of Aristolochia serpentaria and of Aristolochia reticulata.

Natural Order.—Aristolochiaceæ. *Habitat.*—
United States.

Used in tinctura cinchonæ composita.

Officinal Preparations.

Extractum Serpentariæ Fluidum, . . ♏x–f℥ss.
Tinctura Serpentariæ, ... f℥ss–ij.

Officinal Name, CASCARILLA. *English Name*,
CASCARILLA.

Definition.—The *bark* of Croton eluteria.
Natural Order.—Euphorbiaceæ. *Habitat.*—West
Indies.

Used mostly in infusion.

FAMILY II.—EMETICS.

Emetics include those drugs which cause emesis or
vomiting.

VEGETABLE EMETICS.

Officinal Name, IPECACUANHA. *English Name*,
IPECAC.

Definition.—The *root* of Cephaëlis ipecacuanha.
Natural Order.—Rubiaceæ. *Habitat.*—Brazil.

Its activity is due to the alkaloid **emetine.**

**Emetic dose of ipecac in powder, gr. x–xxx; as
diaphoretic and expectorant, gr. ⅛–j.**

Officinal Preparations.

Syrupus Ipecacuanhæ (used almost en-
    tirely for children), emetic dose, f℥j–
    iv ; diaphoretic and expectorant, . . ♏iij–xx.

Vinum Ipecacuanhæ, emetic dose, . . f℥j–ij.
Extractum Ipecacuanhæ Fluidum, eme-
tic dose, ... ℳxx–xl.
Trochisci Ipecacuanhæ, ... ı=gr. ¼ of ipecac.
Tinctura Ipecacuanhæ et Opii, . . . ⎫
Pulvis Ipecacuanhæ et Opii, ... ⎬ See Opium.
Trochisci Morphinæ et Ipecacuanhæ, ⎭

Officinal Name, SANGUINARIA. *English Name*, BLOOD-
ROOT.

Definition.—The *rhizome* of Sanguinaria canadensis.

Natural Order.—Papaveraceæ. *Habitat.*—Widely
diffused.

**Contains the alkaloids chelerythrine (most abun-
dant),** homochelidonine, **sanguinarine,** and proto-
pine.

Dose, as emetic, gr. x–xxx ; rarely used in crude
form.

Officinal Preparations.

Extractum Sanguinariæ Fluidum, . . . ℳj–v.
Tinctura Sanguinariæ, ... ℳxv–xxx.

Officinal Name, APOMORPHINÆ HYDROCHLORAS.
English Name, HYDROCHLORATE OF APOMORPHINE.

Definition.—The hydrochlorate of an artificial
alkaloid prepared from morphine or codeine.

Dose, emetic,—generally given hypodermically,
—gr. $\frac{1}{13}$–⅛ ; expectorant, gr. $\frac{1}{20}$–$\frac{1}{10}$.

Squill is occasionally used as an emetic.

Mustard Flour—the ordinary mustard of the grocer—is one of the emetics most frequently employed. It is given in warm water, about ℥ij to Oj, repeated if necessary in two to five minutes.

MINERAL EMETICS.

Tartar Emetic,—very slow and depressing, but extremely persistent. Not much used.

Dose, gr. ss–ij.

Sulphate of zinc acts promptly, is purely mechanical, and produces no irritation. It acts well in combination with ipecac, say 30 grains of sulphate of zinc with 50 grains of ipecac, and then one-half of the above mixture every fifteen minutes to effect desired result.

Sulphate of copper is more irritating than sulphate of zinc, which is preferable in every way.

Dose, gr. v–x, not to be repeated.

Alum (powdered) has been used in ℥j dose in syrup or molasses for children (as in membranous croup), but is considered unreliable by Dr. Wood.

FAMILY III.—CATHARTICS.

Purgatives or cathartics are drugs producing purgation or catharsis by increasing intestinal secretion or peristaltic action.

They include (1) laxatives, (2) purges, (3) hydragogues, and (4) drastics.

Laxatives simply cause a mild evacuation of the

18

bowels, and do not produce purgation even when given in large doses.

Officinal Name, TAMARINDUS. *English Name*, TAMA-
RIND.

Definition.—The preserved *pulp* of the *fruit* of Tamarindus indica.

Natural Order.—Leguminosæ. *Habitat.*—East and West Indies.

Contains citric, malic, and tartaric acids.

Dose, ℥ss–j, as laxative.

Used in confection of senna.

Officinal Name, MANNA.

Definition.—The *concrete, saccharine exudation* of Fraxinus ornus.

Natural Order.—Oleaceæ. *Habitat.*—Sicily.

Contains *mannite*, an active crystalline principle.

Dose, for adult, ℥ss–ij ; child, ʒj–℥ss.

Officinal Name, CASSIA FISTULA. *English Name*,
PURGING CASSIA.

Definition.—The *fruit* of Cassia fistula.

Natural Order.—Leguminosæ. *Habitat.*—North America.

Dose, ℥j–ij.

Used in confection of senna.

Officinal Name, FRANGULA. *English Name*, BUCK-
THORN.

Definition.—The *bark* of Rhamnus frangula, collected at least one year before being used.

Contains *franguline*, an active principle, and *emodine* (glucoside).

Officinal Preparation.
Extractum Frangulæ Fluidum, ... f℥ ss–ij.

The *bark* of Rhamnus purshiana, or California buckthorn (**cascara sagrada**) is much oftener used than the above.
Officinal Preparation.
Extractum Rhamni Purshianæ Fluid-
um, ... ♏x–f℥j.

Officinal Name, EUONYMUS. *English Name*, WAHOO.

Definition.—The *bark* of the *root* of Euonymus atropurpureus.

Natural Order.—Celastrineæ. *Habitat.*—United States.

Contains *euonymin*,* a bitter principle.

Dose, gr. ij–iv.

Officinal Preparation.
Extractum Euonymi, ... gr. ij–vj.

Officinal Name,	*English Name,*
MAGNESIA.	LIGHT MAGNESIA.
MAGNESIA PONDEROSA.	HEAVY MAGNESIA.

Definition.—These differ only in physical characteristics.

Dose, gr. x–ʒij.

---

* Not officinal.

Officinal Preparations.

Magnesii Carbonas, ... gr. x-ʒ ij.

Magnesii Citras Effervescens, ... gr. x-ʒ ij.

Liquor Magnesii Citratis, ... f ʒ iv–viij.

Magnesii Sulphas (*Epsom salt*), more
active than the above, ... ʒ ij–ʒ j.

SULPHUR. BRIMSTONE.*

Dose, gr. x–xx, t. i. d.

Officinal Preparations.

Sulphur Sublimatum (flowers of sulphur), often given in molasses, . . . ʒ j–iv.

Unguentum Sulphuris.

Sulphur Lotum (washed sulphur), . . ʒ j–iv.

Sulphuris Iodidum.

Sulphur Præcipitatum (precipitated sul-
.phur), ... ʒ j–iv.

Calx sulphurata et potassa sulphurata—sulphurated
lime and sulphurated potassium—are occasionally,
though rarely, used.

Dose, gr. $\frac{1}{10}$–$\frac{1}{4}$.

Purges.—Medicines which cause brisk catharsis
but are not poisonous even in large doses.

Officinal Name, OLEUM RICINI. *English Name*, CASTOR
OIL.

Definition.—A cold-expressed oil from the *seeds*
of Ricinus communis.

* Not officinal.

Natural Order.—Euphorbiaceæ. *Habitat.*—India.

The seeds contain *ricin*, an active poisonous principle which, however, is not communicated to the oil.

Dose, fʒij–fʒj.

Officinal Name, HYDRARGYRUM.   *English Name,*
MERCURY.

Blue mass and calomel are used as cathartics. (See Mercury.)

Officinal Name, RHEUM.   *English Name,* RHUBARB.

Definition.—The *root* of Rheum officinale.

Natural Order.—Polygonaceæ. *Habitat.*—China.

Dose, in powder as stomachic, **gr. j–v**; cathartic, gr. xx–ʒj.

Officinal Preparations.

Extractum rhei, ... **gr. v–x.**
Extractum rhei fluidum, ... ♏x–xxx.
Tinctura rhei, ... ⎫
Tinctura rhei aromatica, ... ⎬ fʒ ss–ij.
Tinctura rhei dulcis, ... ⎭
Mistura rhei et sodæ, ... fℨ ss–ij.
Syrupus rhei (for infants), ... fʒj.
Syrupus rhei aromaticus (for infants), . fʒj.
Pilulæ rhei, ... āā gr. iij.
Pilulæ rhei compositæ (rhubarb, gr. ij; aloes, gr. jss).
Pulvis rhei compositus (ginger, 1; rhubarb, 2; magnesia, 6).

Officinal Name, JUGLANS. *English Name*, BUTTERNUT.

Definition.—The *bark* of the *root* of Juglans cinerea, collected in the autumn.

Natural Order.—Juglandaceæ. *Habitat.*—United States.

Officinal Preparation.

Extractum Juglandis, ... **gr. x–xxx.**

Officinal Name, ALOE (Pharm., 1880), ALOE BARBADEN-SIS, ALOE SOCOTRINA. *English Name*, ALOES.

Definition.—The *inspissated juice* of the *leaves* of Aloe vera (Barbadoes aloes) and Aloe Perriyi (socotrine aloes).

Natural Order.—Liliaceæ. *Habitat.*—Socotra, Zanzibar.

Contains *aloinum* (*aloin*), a neutral principle.

Officinal Preparations.

Aloe Purificata, ... **gr. ss–x.**
Extractum Aloes, ... **gr. ss–v.**
Pilulæ Aloes, soap and aloes, . . **aa gr. ij.**
Pilulæ Aloes et Asafœtidæ, aloes, asafe-
   tida, and soap, of each, 1⅓ grs.
Pilulæ Aloes et Mastiches (*Lady Web-
   ster's pills*), aloes, 2 grs. ; mastic and
   rose, of each, ½ gr.
Pilulæ Aloes et Ferri, aloes and ferrous
   sulphate, of each, 1 gr.
Pilulæ Aloes et Myrrhæ, aloes, 2 grs. ;
   myrrh, 1 gr. ; aromatic powder, ½ gr.
Tinctura Aloes, ... **f ℥ ss–ij.**
Tinctura Aloes et Myrrhæ, . . . **f ℥ ss–ij.**
Aloinum, ... **gr. 1/10–j.**

Officinal Name, SENNA. *English Name*, SENNA.

Definition.—The *leaflets* of Cassia acutifolia and Cassia angustifolia.

Natural Order.—Leguminosæ. *Habitat.*—Egypt and Arabia.

Officinal Preparations.

Confectio Sennæ, ... ʒ j–iv.
Syrupus Sennæ, ... f ʒ j–iv.
Extractum Sennæ Fluidum, ... f ʒ j–iv.
Infusum Sennæ Compositum (black
draught), ... f ʒ ij–f ℥ ss.

Used also in pulvis glycyrrhizæ compositus (compound liquorice powder).

Hydragogues "(including the **salines**) produce large watery stools without much irritation."

Officinal Preparations of Magnesia used as Purgatives.

Magnesii Sulphas (Epsom salt), . . . gr. x–ʒ j.
Magnesii Citras Effervescens, ... gr. x–ʒ j.
Liquor Magnesii Citratis, ... f ℥ ij–iv.

Officinal Name, SODII SULPHAS. *English Name*,
GLAUBER'S SALT.

Definition.—Not much used ; practically the same as sulphate of magnesia, but of more disagreeable taste.

Officinal Name, SODII PHOSPHAS. *English Name*,
PHOSPHATE OF SODIUM.

Dose, gr. x–ʒiv.

Officinal Name, POTASSII ET SODII TARTRAS.
English Name, ROCHELLE SALT.

Definition.—Occurs in large crystals, but usually kept in powdered form.

Dose, ʒss–ij, in water.

Officinal Name, PULVIS EFFERVESCENS COMPOSITUS.
English Name, SEIDLITZ POWDER.

Definition.—Contains two papers: the white is made of 35 grains of tartaric acid, the blue of 40 grains of sodium bicarbonate and 120 grains of Rochelle salt.

Dose, one powder. Each paper to be dissolved in a separate glass of water, mix, and drink while effervescing.

Drastics.—Active, irritant vegetable cathartics; in sufficient amount may cause death.

Officinal Name, JALAPA. *English Name*, JALAP.

Definition.—The *tuberous root* of Ipomœa jalapa.
Natural Order. — Convolvulaceæ. *Habitat.* — Mexico.

Contains an active resin, to which its properties are due.

Officinal Preparations.

Extractum Jalapæ, ... gr. iij–x.
Resina Jalapæ, ...gr. ij–v.
Pulvis Jalapæ Compositus, jalap, 35
   per cent. ; potassium bitartrate, 65 per
   cent., ... gr. x–ʒj.

Officinal Name, COLOCYNTHIS. *English Name*,
COLOCYNTH.

Definition.—The *fruit* of Citrullus colocynthis, *deprived of its rind.*

Natural Order.—Cucurbitaceæ. *Habitat.*—Africa.

Officinal Preparations.

Extractum Colocynthidis, ... gr. ij–v.
Extractum Colocynthidis Compositum,
laxative dose, gr. j–iij; purgative
dose, ... gr. x–xx.
Pilulæ Catharticæ Compositæ, contain compound ext. of colocynth, 80
gm. ; mild mercurous chloride, 60
gm.; ext. of jalap, 30 gm. ; gamboge,
15 gm.; water, q. s., to make 1000
pills, ... j–iij pills.

Officinal Name, SCAMMONIUM. *English Name*,
SCAMMONY.

Definition.—A *resinous exudation* from the living *root* of Convolvulus scammonia.

Natural Order. — Convolvulaceæ. *Habitat.* —
Syria.

Its activity is due to the resin *scammonin.*

Dose, gr. v–xx.

Officinal Preparation.

Resina Scammonii, ... gr. ij–x.

19

Officinal Name, PODOPHYLLUM. *English Name*,
MAY APPLE.

Definition.—The *rhizome* and *roots* of Podophyllum peltatum.

Natural Order.—Berberideæ. *Habitat.*—United States.

Contains the alkaloid berberine and several resins, to which its activity is due.

Dose, in powder, gr. x–xx.

Officinal Preparations.

Extractum Podophylli, ... gr. v–x.
Extractum Podophylli Fluidum, . . . ℳ x–xx.
Resina Podophylli (podophyllin), . . gr. ⅙–¼.

Officinal Name, ELATERINUM. *English Name*, ELATERIN.

Definition.—A neutral principle obtained from *elaterium*, a substance deposited by the *juice* of Ecballium elaterium (squirting cucumber).

Natural Order.—Cucurbitaceæ. *Habitat.*—Europe.

Dose, gr. $\frac{1}{12}$–$\frac{1}{6}$.

Officinal Preparation.

Trituratio Elaterini, ... gr. ¼–j.

Officinal Name, CAMBOGIA. *English Name*, GAMBOGE.

Definition.—A *gum resin* obtained from Garcinia Hanburii.

Natural Order.—Guttiferæ. *Habitat.*—Siam.
Used in pilulæ catharticæ compositæ.

Officinal Name, OLEUM TIGLII. *English Name*,
CROTON OIL.

Definition.—A *fixed oil* expressed from the *seed*
of Croton tiglium.
Natural Order.—Euphorbiaceæ. *Habitat.*—India.
Dose, ℥j–ij.
Used externally as counter-irritant.

FAMILY IV.—DIURETICS.

Medicines which increase the flow of urine. They
include (1) the hydragogue diuretics, which simply
increase the flow of water from the kidneys, and are
therefore useful in dropsy. (2) Refrigerant diuretics,
which exert a marked sedative action, and so modify
the secretion that they render the urine less irritant.
(3) Alterative diuretics, whose active principles are .
eliminated by the kidneys and thus act on the mucous
surfaces over which they pass.

HYDRAGOGUE DIURETICS.

Officinal Name, SCILLA. *English Name*, SQUILL.

Definition.—The sliced *bulb* of Urginea mari-
tima.
Natural Order. — Liliaceæ. *Habitat.* — South
Europe.
Dose, in powder, gr. j–ij.

Officinal Preparations.

Acetum Scillæ, ... ℥x–f℥ss.
Extractum Scillæ Fluidum, ... ℥j–ij.
Tinctura Scillæ, ... ℥x–f℥ss.
Syrupus Scillæ, ... f℥ss–j.
Syrupus Scillæ Compositus (*Cox's hive*
　　syrup), 2 parts in 1000 of tartar emetic, ℥v–xxx.

Officinal Name, SCOPARIUS.　　*English Name*, BROOM.

Definition.—The *tops* of Cytisus scoparius.
Natural Order.—Leguminosæ. *Habitat.*—Europe.
Contains the alkaloid **sparteine** and scoparin.

Officinal Preparation.

Extractum Scoparii Fluidum, ... ℥ij–x.

Calomel in large doses is occasionally used as a
diuretic.

The alkaloid **theobromine**,* used as the salicylate
or double salicylate of sodium and theobromine (di-
uretine), in daily doses of gr. xv–c, is employed at
times, and is a valuable diuretic.

BLATTA.*

Definition.—The dried bodies of Blatta orien-
talis (cockroach) have often been used as a popular
remedy in dropsy.

Dose, gr. xv–xxx in twenty-four hours.

* Not officinal.

Officinal Name, SPIRITUS ÆTHERIS NITROSI. *English Name*, SPIRIT OF NITROUS ETHER.

Definition.—An alcoholic solution of ethyl nitrite, yielding not less than 11 times its own volume of nitrogen dioxid.

Dose, as diuretic, fȝj–iv.

Caffeine, jaborandi, strophanthus, and digitalis are all valuable diuretics (see Doses, etc., under proper headings).

REFRIGERANT DIURETICS.

POTASSIUM.*

Definition.—Obtained from vegetable *ash, nitre,* and from the *argol* or tartar deposited by wine.

Officinal Preparations.

Potassa (caustic).

Liquor Potassæ (five per cent. potassium
    hydrate), ... ♏v–xx.

Liquor Potassii Citratis, citric acid, six
    parts ; potass. bicarb., eight parts in
    100, ... fȝss.

Potassii Acetas, ... ⎫
Potassii Chloras, ... ⎪
Potassii Citras, ... ⎪
Potassii Carbonas, ... ⎬ gr. v–xxx.
Potassii Bicarbonas, ... ⎪
Potassii Nitras (saltpetre), . . . ⎪
Potassii Bitartras (*cream of tartar*), ⎭

---

* Not official.

Potassii et Sodii Tartras (Rochelle salt), ℥j.
Potassii Sulphas, ... ℥j.
Trochisci Potassii Chloratis, ... I = gr. v.
Liquor Potassii Citratis (Citric acid
neutralized by potass. bicarb.), . . f℥ss-j.
Charta Potassii Nitratis, made from the nitrate (nitre).
Potassii Bichromas.
Potassii Ferrocyanidum.
Potassa Sulphurata.
Potassa cum Calce (caustic).

Officinal Name, LITHII CARBONAS. *English Name*,
LITHIUM CARBONATE.

Dose, gr. v–xv, t. i. d.

Besides the carbonate, which is diuretic, there are
officinal:

Lithii Salicylas, ... :
Lithii Bromidum, ...
Lithii Benzoas, ... } gr. v–xx.
Lithii Citras, ...

PIPERAZINE.*

Definition.—Occurs in small, glossy crystals.
Its value depends entirely on its solvent power over
uric acid.

May be given hypodermically in two per cent.
solution, or gr. xv, in twenty-four hours, in water,
as it is highly hygroscopic.

---

* Not officinal.

Strontium iodide, lactate, and bromide, the salts of strontium, are sometimes employed.

Dose, gr. xxx in twenty-four hours.

ALTERATIVE DIURETICS.

Officinal Name, BUCHU. *English Name*, BUCHU.

Definition.—The *leaves* of Barosma betulina and Barosma crenulata.

Natural Order.—Rutaceæ. *Habitat.*—Africa.

Officinal Preparation.

Extractum Buchu Fluidum, ... ℥xx–f℥j.

Officinal Name, PAREIRA. *English Name*, PAREIRA BRAVA.

Definition.—The *root* of Chondodendron tomentosum.

Natural Order.—Menispermaceæ. *Habitat.*—Brazil.

Officinal Preparation.

Extractum Pareiræ Fluidum, ... f℥ss–ij.

Officinal Name, UVA URSI. *English Name*, BEARBERRY.

Definition.—The *leaves* of Arctostaphylos uva-ursi.

Natural Order.—Ericaceæ. *Habitat.*—United States and Europe.

Contains the glucoside *arbutin*, to which its activity is due.

Officinal Preparations.

Extractum Uvæ Ursi, ... gr. x–xv.
Extractum Uvæ Ursi Fluidum, ... f℥ss–ij.

Officinal Name, CHIMAPHILA.   *English Name*, PIPSIS-
SEWA.

Definition.—The *leaves* of Chimaphila umbellata.
Natural Order.—Ericaceæ.   *Habitat.*—United
States.

Officinal Preparation.

Extractum Chimaphilæ Fluidum, . . f℥ss–j.

Officinal Name, JUNIPERUS (Pharm., 1880).   *English
Name*, JUNIPER.

Definition.—The *fruit* of Juniperus communis.
Natural Order.—Coniferæ.   *Habitat.*—United
States and Europe.
Contains a volatile oil.

Officinal Preparations.

Oleum Juniperi, ... ♏ij–v.
Spiritus Juniperi (five per cent. of oil), f℥ss–ij.
Spiritus Juniperi Compositus (contains
 oils of juniper, caraway, and fennel), f℥j–ij.

Officinal Name, OLEUM ERIGERONTIS.   *English Name*,
OIL OF ERIGERON.

Definition.—A volatile oil distilled from the
fresh flowering herb of Erigeron canadense or Canada
fleabane.

Officinal Name, TEREBINTHINA CANADENSIS. *English Name*, CANADA TURPENTINE, CANADA BALSAM, BALSAM OF FIR.

Definition.—A liquid *oleoresin* from Abies balsamea.

Rarely used.

Officinal Name, COPAIBA. *English Name*, COPAIBA.

Definition.—The *oleoresin* of Copaiba Langsdorffii.

Natural Order.—Leguminosæ. *Habitat.*—South America.

Dose, ℞xx, t. i. d.

Officinal Preparations.

Massa Copaibæ, ...	℞v–xx.
Oleum Copaibæ, ...	gr. v–xv.
Resina Copaibæ, ...	gr. v–xv.

Officinal Name, CUBEBA. *English Name*, CUBEBS.

Definition.—The unripe *fruit* of Piper cubeba.

Natural Order.—Piperaceæ. *Habitat.*—East Indies.

Contains cubebic acid, a volatile oil, and the neutral principle *cubebin.*

Dose, in powder, ʒss–j.

Officinal Preparations.

Oleum Cubebæ, ...	℞v–xv.
Tinctura Cubebæ, ...	fʒss–ij.
Extractum Cubebæ Fluidum, ...	℞x–xxx.
Oleoresina Cubebæ, ...	℞v–xv.

Officinal Name, MATICO. *English Name*, MATICO.

Definition.—The *leaves* of Piper angustifolium.
Natural Order.—Piperaceæ. *Habitat.*—Peru.
Contains a volatile oil, resin, and bitter principle.

Officinal Preparations.

Extractum Matico Fluidum, ... f℥ ss–j.
Tinctura Matico, ... f℥ ss–ij.

FAMILY V.—DIAPHORETICS.

Medicines which increase the flow of perspiration,
acting on the skin directly or through the system ;
they include : (1) Nauseating diaphoretics, (2) refrigerant diaphoretics, and (3) simple diaphoretics.

The **nauseating diaphoretics** include tartar
emetic, ipecac and its preparations, notably Dover's
powder.

The **refrigerant diaphoretics** include aconite,
veratrum viride, the cardiac depressants, and, best
of all, potassium citras—either as neutral mixture
or effervescing draught.

The **simple diaphoretics** include:

Officinal Name, PILOCARPUS. *English Name*,
JABORANDI.

Definition.—The *leaflets* of Pilocarpus selloanus
and Pilocarpus jaborandi.
Natural Order.—Rutaceæ. *Habitat.*—Brazil.
Contains the alkaloid, pilocarpine.
Dose, of crude drug, gr. v to xl.

Officinal Preparations.

Extractum Pilocarpi Fluidum, ... ℥v-l.

Pilocarpinæ Hydrochloras, ... gr. ⅛-¼.

Officinal Name, LIQUOR AMMONII ACETATIS. *English Name,* SPIRIT OF MINDERERUS.

Definition.—Dilute acetic acid, neutralized by carbonate of ammonium. Valuable as a basis for fever mixtures.

Dose, f℥j-f℥ss.

Officinal Name, SPIRITUS ÆTHERIS NITROSI. *English Name,* SPIRIT OF NITROUS ETHER.

Definition.—An alcoholic solution of ethyl nitrite yielding not less than 11 times its own volume of nitrogen dioxid.

Dose, as diaphoretic, f℥j-f℥ss.

FAMILY VI.—EXPECTORANTS.

Remedies which cause an increase or modification in the amount of secretion from the larger tubes of the respiratory tract, and facilitate the expulsion thereof.

The **nauseating expectorants** are lobelia, tartar emetic, and ipecac. (See doses elsewhere.)

Officinal Name, GRINDELIA. *English Name,* GRINDELIA.

Definition.—The *leaves* and *flowering tops* of Grindelia robusta and Grindelia squarrosa.

Natural Order.—Compositæ. *Habitat.*—United States.

Officinal Preparation.

Extractum Grindeliæ Fluidum, . . . ♏x-f℥ij.

STIMULATING EXPECTORANTS.

Ammonium chloride (ammonii chloridum), in doses of gr. v-xx, is a valuable expectorant.

Officinal Name, SENEGA. *English Name,* SENEGA.

Definition.—The *root* of Polygala senega.
Natural Order.—Polygaleæ. *Habitat.*—United States.
Contains the glucoside *senegin,* and polygallic acid.

Officinal Preparations.

Extractum Senegæ Fluidum, ... ♏x-xxx.
Syrupus Senegæ, ... f℥ss-ij.

Officinal Name, AMMONIACUM. *English Name,* AMMONIAC.

Definition.—A *gum resin* from Dorema ammoniacum.
Natural Order.—Umbelliferæ. *Habitat.*—Persia.
Dose, gr. x-xx.

Officinal Preparations.

Emplastrum Ammoniaci cum Hydrargyro.
Emulsum Ammoniaci, ... f℥ss-j.

SULPHURETTED HYDROGEN. *

Rarely used.

Officinal Name, BENZOINUM.   *English Name*, BENZOIN.

Definition.—A *balsamic resin* obtained from Styrax benzoin.

Natural Order.—Styraceæ.   *Habitat.*—Peru.

Contains *benzoic acid*, a volatile oil, and a resin.

Officinal Preparations.

Adeps Benzoinatus, ... external use.
Tinctura Benzoini, ... f ℨ ss–ij.
Tinctura Benzoini Composita, ... f ℨ ss–ij.
Acidum Benzoicum, ... gr. x– ℨ ss.
Ammonii,
Lithii,   } Benzoas, ... gr. x–xv.
Sodii,

Officinal Name, BALSAMUM PERUVIANUM.   *English Name*, BALSAM OF PERU.

Definition.—A *balsam* from Toluifera pereiræ.

Natural Order.—Leguminosæ.   *Habitat.*—South America.

Dose, ℨss.

Officinal Name, BALSAMUM TOLUTANUM.   *English Name*, BALSAM OF TOLU.

Definition.—A *balsam* obtained from Toluifera balsamum.

---

* Not officinal.

Natural Order.—Leguminosæ. *Habitat.*—Central America.

Officinal Preparations.

Tinctura Tolutana, ... f℥ j–iij.
Syrupus Tolutanus, ... f℥ j–ij.

Also used in the compound tincture of benzoin. Mainly used as vehicles.

Officinal Name, MYRRHA. *English Name*, MYRRH.

Definition.—A *gum-resin* from Commiphora myrrha.

Natural Order.—Burseraceæ. *Habitat.*—Arabia, Africa.

Officinal Preparations.

Pilulæ Aloes et Myrrhæ, ... j–iij pills.
Tinctura Aloes et Myrrhæ, ... } f℥ ss–ij.
Tinctura Myrrhæ, ... }
Mistura Ferri Composita, ... f℥ ss.

Officinal Name, ALLIUM. *English Name*, GARLIC.

Definition.—The *bulb* of Allium sativum.

Natural Order.—Liliaceæ. *Habitat.*—Indigenous.

Officinal Preparation.

Syrupus Allii, ... f℥ j–ij.

Syrup of squill and compound syrup of squill are also sometimes used as stimulating expectorants in doses of f℥ss.

Officinal Name, PIX LIQUIDA. *English Name*, TAR.

Definition.—The empyreumatic *oleoresin* obtained by destructive distillation of the *wood* of Pinus palustris and other species of Pinus.

Natural Order.—Coniferæ. *Habitat.*—Indigenous.

On distillation it yields pitch, oil of tar, and pyroligneous acid.

Officinal Preparations.

Oleum Picis Liquidæ, ... external use.
Syrupus Picis Liquidæ, ... f℥j–iv.
Unguentum Picis Liquidæ, 50 per cent.
    tar.

Officinal Name, TEREBENUM. *English Name*,
TEREBENE.

Definition.—A clear, colorless *fluid*, obtained by the action of sulphuric acid on oil of turpentine, and then distilled.

Consists mostly of pinene and very small amounts of terpinene and dipentene.

Dose, ♏x, in capsule or emulsion, t. i. d.

FAMILY VII.—EMMENAGOGUES.

Remedies used to increase or re-establish the menstrual flow.

They include tonic and stimulating emmenagogues.

The tonic emmenagogues are *iron, myrrh,* and *aloes.* Iron and myrrh act simply by their general

tonic action and are largely used in anemic affections of the menstrual flow.   The effect of aloes is
due solely to stimulation of the rectum.

STIMULATING EMMENAGOGUES.

Officinal Name, SABINA.   *English Name*, SAVINE.

Definition.—The *tops* of Juniperus sabina.
Natural Order.—Coniferæ.   *Habitat.*—Europe.
Contains a *volatile oil*, to which its activity is due.

Officinal Preparations.

Oleum Sabinæ, ... ℥v–x.
Extractum Sabinæ Fluidum, ... ℥v–xv.

RUTA.   RUE.*

Definition.—The *leaves* of Ruta graveolens.
Natural Order.—Rutaceæ.   *Habitat.*—Europe.
Contains a *volatile oil*.

Preparation.

Oleum Rutæ, ... ℥ij–vj.

APIOL.*

Definition.—A liquid, neutral principle from
Petroselinum sativum—ordinary parsley.
Natural Order.—Umbelliferæ.
Dose, ℥iij–x, in capsule.

Potassii permanganas, in doses of gr. j–ij,
t. i. d. ; **cantharides,** in the form of the tincture,

---

* Not official.

♏ iij–v, and the ammoniated tincture of guaiac, in doses of ʒss–j, have often been employed with success.

Officinal Name, TANACETUM.   *English Name*, TANSY.

Definition.—The *leaves* and *tops* of Tanacetum vulgare.

Natural Order.—Compositæ.   *Habitat.*—Indigenons.

Contains a *volatile oil* and bitter principle.

Dose, gr. x–xx. The oil is dangerous, dose ♏ j–v.

Officinal Name, HEDEOMA.   *English Name*, PENNY-
ROYAL.

Definition.—The *leaves* and *tops* of Hedeoma pulegioides.

Natural Order. — Labiatæ.   *Habitat.*— United States.

Owes its activity to a *volatile oil.*

Officinal Preparation.

Oleum Hedeomæ, ... ♏ j–v.

FAMILY VIII.—OXYTOCICS.

Remedies which increase uterine muscular contraction.

Officinal Name, ERGOTA.   *English Name*, ERGOT.

Definition.—The *sclerotium* of Claviceps purpurea (*natural order*, Fungi), replacing the grain of

rye (Secale cereale, *natural order*, Gramineæ), a fungous growth from the diseased ovary of the rye. Should not be over one year old.

Dose, in powder, gr. xxx.

Officinal Preparations.

Extractum Ergotæ Fluidum, ... ſʒ ss–j.
Vinum Ergotæ, ... ſʒ ij–ſʒ ss.

Ergotin, a concentrated extract of ergot, is sometimes used ; gr. v equal to ſʒj of fluid extract.

Hydrastis and hydrastinine hydrochlorate have lately been used as oxytocics with good results. (See doses elsewhere.)

Officinal Name, GOSSYPII RADICIS CORTEX. *English Name*, COTTON ROOT BARK.

Definition.—The *bark* of the *root* of Gossypium herbaceum.

Natural Order.—Malvaceæ. *Habitat.*—United States.

Used in decoction (ʒiv to Oj of water) fʒij, repeated as needed.

Officinal Preparation.

Extractum Gossypii Radicis Fluidum, ſʒ ss- ij.

USTILAGO. SMUT OF INDIAN CORN.*

Definition.—Ustilago maydis (*natural order,*

---

* Not officinal.

Fungi), corn smut or corn ergot, a fungous growth on Zea mays (Indian corn).

Natural Order.—Gramineæ.

Dose, gr. xv–ʒj.

FAMILY IX.—SIALAGOGUES.

Those remedies which increase the flow of saliva and oral mucus.

Officinal Name, PYRETHRUM.   *English Name*, PELLITORY.

Definition.—The *root* of Anacyclus pyrethrum.

Natural Order.—Compositæ.   *Habitat.*—Europe.

Dose, ʒss–j to be chewed ; or Tinctura Pyrethri, fʒss–ij.

FAMILY X.—ERRHINES.

Remedies acting on the nasal mucous membrane.

FAMILY XI.—EPISPASTICS.

Drugs used to produce blisters.

Officinal Name, CANTHARIS.   *English Name*, CANTHARIDES, SPANISH FLIES.

Definition.—A beetle, Cantharis vesicatoria, inhabiting Spain, Italy, and Southern Europe.

Order.—Coleoptera.

The dried bodies contain a volatile oil and a neutral principle, *cantharidin*, to which is due the vesicating property.

Officinal Preparations.

Ceratum Cantharidis, ... external use.
Tinctura Cantharidis, ... ℥ij–v.
Collodium Cantharidatum, for blistering.
Emplastrum Picis Cantharidatum, warm-
    ing plaster.

FAMILY XII.—RUBEFACIENTS.

Remedies causing powerful irritation and conges-
tion of the skin surface, which is, however, of short
duration.

Officinal Name,	*English Name,*
SINAPIS ALBA.	WHITE MUSTARD.
SINAPIS NIGRA.	BLACK MUSTARD.

Definition.—The *seed* of Brassica alba (white),
and Brassica nigra (black),—mustard.

Natural Order.—Cruciferæ.   *Habitat.*—Europe.

Officinal Preparations (from black mustard).

Charta Sinapis, ... 6 grs. to square in.
Oleum Sinapis, ... ℥ij–v, diluted.
Linimentum Sinapis Compositum.

Capsicum, oleum terebinthinæ, and ammonia are also employed as rubefacients.

Officinal Name, PIX BURGUNDICA. *English Name,*
BURGUNDY PITCH.

Definition.—The prepared *resinous exudation* of
Abies excelsa, or Norway spruce.

Natural Order.—Coniferæ.

Contains a resin and a volatile oil.

Officinal Preparations.
Emplastrum Picis·Burgundicæ.
Emplastrum Picis Cantharidatum, warming plaster.

FAMILY XIII.—ESCHAROTICS.

Those remedies which, by contact, destroy either healthy or diseased tissues.

Officinal Name, ACIDUM CHROMICUM.   *English Name*, CHROMIC ACID.

Definition.—Acicular crystals of deep-red color, very deliquescent, made by the action of sulphuric acid on potassium bichromate.

Apply with glass rod or platinum wire.

Officinal Name, BROMUM.   *English Name*, BROMINE.

Definition.—A heavy, dark-red, mobile liquid ; very powerful caustic.

Besides the above are used :

Alumen Exsiccatum.
Acidum Sulphuricum.
  "   Nitricum.
  "   Hydrochloricum.
  "   Arsenosum.
Hydrargyri Chloridum Corrosivum.
Cupri Sulphas.
Zinci Sulphas.
Zinci Chloridum.
Potassa (caustic potash).
Argenti Nitras Fusus (lunar caustic).
Potassa cum Calce (Vienna paste).
Liquor Potassæ, etc.

FAMILY XIV.—DEMULCENTS.

Bland substances, capable of soothing inflamed surfaces; mostly of a gummy or mucilaginous consistency when mixed with water.

Officinal Name, ACACIA. *English Name*, GUM ARABIC.

Definition.—A gummy exudation from Acacia senegal.

Natural Order.—Leguminosæ. *Habitat.*—Africa and Australia.

Officinal Preparations.

Mucilago Acaciæ, ⎫
Syrupus Acaciæ, ⎭ . . . . vehicles.

Officinal Name, TRAGACANTHA. *English Name*, TRAGACANTH.

Definition.—A *gummy exudation* from Astragalus gummifer.

Natural Order.—Leguminosæ. *Habitat.*—Asia Minor.

Officinal Preparation.

Mucilago Tragacanthæ, ... vehicle.

Officinal Name, ULMUS. *English Name*, SLIPPERY ELM.

Definition.—The *inner bark* of Ulmus fulva.

Natural Order.—Urticaceæ. *Habitat.*—United States.

Officinal Preparation.

Mucilago Ulmi, ... vehicle.

Officinal Name, CETRARIA.   *English Name*, ICELAND
MOSS.

Definition.—A *lichen*, Cetraria islandica, found
in Iceland.

Natural Order.—Lichenes.

Officinal Preparation.

Decoctum Cetrariæ, ... f℥ss–iv.

· *Officinal Name*, CHONDRUS.   *English Names*, IRISH
MOSS, CARRAGHEEN.

Definition.—*Fronds* of Chondrus crispus and
Gigartina mamillosa.

Natural Order.—Algæ.   *Habitat.*—New England
and Ireland.

Nutrient and demulcent.

Officinal Name, GLYCYRRHIZA.   *English Name*,
LIQUORICE ROOT.

Definition.—The *root* of Glycyrrhiza glabra.

Natural Order.—Leguminosæ.   *Habitat.*—Europe.

Contains the glucoside *glycyrrhizin*.

Officinal Preparations.

Extractum Glycyrrhizæ, ... ⎫
Extractum Glycyrrhizæ Fluidum, . ⎬ for flavoring and as
Extractum Glycyrrhizæ Purum, . . ⎭ vehicles.
Glycyrrhizinum Ammoniatum, ... gr. v–x.
Pulvis Glycyrrhizæ Compositus (cathartic),

contains senna, fennel, and washed
sulphur, ... ℥ ss–ij.
Mistura Glycyrrhizæ Composita (Brown
mixture) contains wine of antimony,
paregoric, and sweet spirit of nitre, . f℥ ss–j.
Trochisci Glycyrrhizæ et Opii, 1 = ext.
of liquorice, 2 grs., ext. of opium,
gr. $\frac{1}{20}$.

Officinal Name, LINUM. *English Name*, FLAXSEED.

Definition.—The *seed* of Linum usitatissimum.

Natural Order.—Lineæ. *Habitat.*—Widely diffused.

Used as tea.

Officinal Preparation.

Oleum Lini (*flaxseed or linseed oil*).

Officinal Name, SASSAFRAS MEDULLA. *English Name*,
SASSAFRAS PITH.

Definition.—The *pith* of Sassafras variifolium.

Natural Order.—Laurineæ. *Habitat.*—Europe
and United States.

It yields a mucilage much used in the treatment
of eye affections.

Officinal Name, ALTHÆA. *English Name*,
MARSHMALLOW.

Definition.—The *root* of Althæa officinalis.

Natural Order.—Malvaceæ. *Habitat.*—United
States.

Officinal Preparation.

Syrupus Althææ.

22

TAPIOCA.*

Definition.—The *fecula* obtained from the *root* of Janipha manihot.

Habitat.—South America.

Maranta (arrowroot), **sago**, and **hordeum** (barley), used mostly as food, are occasionally employed as demulcents.

FAMILY XV.—EMOLLIENTS.

Bland, fatty substances, which soothe and soften the skin.

Officinal Name, GLYCERINUM. *English Name*, GLYCERIN.

Definition.—A clear, colorless liquid of thick syrupy consistence ; odorless ; very sweet, and slightly warm to the taste. Obtained by the decomposition of vegetable or animal fats or fixed oils, and containing at least 95 per cent. of absolute glycerin.

Officinal Preparation.

Suppositoria Glycerini.

LANOLIN.*

Definition.—Purified fat of sheep's wool. Often used as an ointment base ; more readily absorbed through the skin than most other fats, according to some authorities.

---

* Not officinal.

Officinal Name, ADEPS. *English Name*, LARD.

Definition.—The prepared fat of Sus scrofa (hog), contains olein and stearin.

Officinal Preparations.

Adeps Benzoinatus.
Unguentum, lard, four parts ; yellow wax, one part.
Ceratum, lard, 70 per cent. ; white wax, 30 per cent.

Officinal Name, CETACEUM. *English Name*, SPERMA-
CETI.

Definition.—A concrete fatty substance, obtained from Physeter macrocephalus (whale).

Officinal Preparations.

Ceratum Cetacei.
Unguentum Aquæ Rosæ.

Officinal Name,	*English Name*,
CERA FLAVA.	YELLOW WAX.
CERA ALBA.	WHITE WAX.

Definition.—Beeswax ; prepared by Apis melli-fica (honey-bee).

Officinal Name, OLEUM THEOBROMATIS. *English Name*, BUTTER OF CACAO.

Definition.—A fixed *oil* expressed from the *seed* of Theobroma cacao. Used for suppositories and in ointments.

Natural Order.—Sterculiaceæ. *Habitat.*—South America.

Officinal Name,   *English Name,*
PETROLATUM MOLLE.   SOFT PETROLATUM.
PETROLATUM SPISSUM.   HARD PETROLATUM.

Definition.—A mixture of hydrocarbons obtained by the distillation of petroleum. When "petrolatum" is prescribed, it means always the soft variety.

FAMILY XVI.—DILUENTS.

These are substances (water and medicated waters) which are to be absorbed during their passage through the body, and so dilute its various fluids and excretions.

FAMILY XVII.—PROTECTIVES.

External applications to exclude air and protect inflamed surfaces.

Officinal Name, COLLODIUM.   *English Name,* COLLO-
DION.

Definition.—A solution of pyroxylin or gun-cotton in alcohol and ether. The alcohol and ether evaporate rapidly and leave a translucent, flexible, adherent film on the skin which is impervious to air and water.

Officinal Preparations.

Collodium Flexile, five per cent. Canada turpentine,
    three per cent. castor oil, and 92 per cent. collodion.
Collodium Stypticum, 20 per cent. tannic acid.
Collodium Cantharidatum, blistering.

LIQUOR GUTTA-PERCHA (TRAUMATICINE).*

Contains nine per cent. of gutta-percha in commercial chloroform. Leaves a film of gutta-percha at place of application after evaporation of the chloroform, and in this manner various remedies are occasionally employed, especially in the treatment of skin diseases.

DIVISION II.

EXTRANEOUS REMEDIES.

FAMILY I.—ANTACIDS.

Remedies used to overcome excessive acidity.

SODIUM (Metal).*

Officinal Preparations.

Soda (caustic).
Liquor Sodæ, five per cent. sodium hy-
   drate, ... ♏ij–x.
Sodii Bicarbonas, ... gr. v–xxx.
Sodii Carbonas, ... gr. v–xxx.
Trochisci Sodii Bicarbonatis, . . . ⎫
Sodii Acetas, ... ⎪
Sodii Benzoas, ... ⎪
Sodii Boras (borax), ... ⎪
Sodii Bromidum, ... ⎪
Sodii Chloras, ... ⎬ gr. v–xxx.
Sodii Chloridum (salt), ... ⎪
Sodii Iodidum, ... ⎪
Sodii Phosphas, ... ⎪
Sodii Salicylas, ... ⎪
Sodii Sulphi Carbolas, ... ⎪
Sodii Sulphas (Glauber's salt), . . . ⎭

The liquor, carbonate, and bicarbonate are oftenest used.

---

* Not officinal.

CALCIUM (Metal).*

Officinal Preparations.

Calcii Bromidum, ... }
Calcii Chloridum, ... } gr. x–xxx.
Calcii Phosphas Præcipitatus, . . . }
Calcii Hypophosphis, ... gr. iij–v.
Calcii Carbonas Præcipitatus.
Calx (quick-lime), caustic.
Liquor Calcis (lime-water).
Linimentum Calcis (*carron oil*), equal
    parts of lime-water and olive oil.
Calx Sulphurata, ... gr. $\frac{1}{10}$–ss.
Creta Præparata (prepared chalk), . . gr. x–$\mathfrak{Z}$j.
Mistura Cretæ, ... f $\mathfrak{Z}$ ss.
Syrupus Calcli Lactophosphatis, . . . f $\mathfrak{Z}$ ss.
Syrupus Calcis, ... m x–f $\mathfrak{Z}$ j.
Syrupus Hypophosphitum, ... f $\mathfrak{Z}$ ij–iv.
Syrupus Hypophosphitum cum Ferro, . f $\mathfrak{Z}$ ij–iv.

FAMILY II.—ANTHELMINTICS.

Remedies causing the expulsion or death of intestinal worms.

Officinal Name, SPIGELIA.   *English Name*, PINKROOT.

Definition.—The *rhizome* and *rootlets* of Spigelia marilandica, or Carolina pink.

Natural Order.—Loganiaceæ.   *Habitat.*—United States.

Officinal Preparation.

Extractum Spigeliæ Fluidum, . . . f $\mathfrak{Z}$ ss–j.

---

* Not officinal.

AZEDARACH.*

Definition.—The *bark* of the *root* of Melia azedarach (pride of China).

Natural Order.—Meliaceæ.

Used in f℥ss doses of decoction, ℥iv to Oj of water boiled to Oij.

Officinal Name, CHENOPODIUM. *English Name*, AMERICAN WORMSEED.

Definition.—The *fruit* of Chenopodium ambrosioides ; variety, anthelminticum.

Natural Order.—Chenopodiaceæ. Its effects are due to the volatile oil it contains.

Dose, of oil, ℳ v–x, for three-year-old child.

Officinal Name, CUSSO (BRAYERA, Pharm., 1880). *English Name*, KOUSSO.

Definition.—The female *inflorescence* of Hagenia abyssinica.

Natural Order.—Rosaceæ. *Habitat.*—Abyssinia.

Officinal Preparation.

Extractum Cusso Fluidum, ... f℥j–iij.

Officinal Name, SANTONICA. *English Name*, LEVANT WORMSEED.

Definition.—The unexpanded *flower-heads* of Artemisia pauciflora.

---

* Not officinal.

Natural Order.—Compositæ. *Habitat.*—Europe and Asia.

Contains the neutral principle *santoninum* (*santonin*), to which its activity is due. Used for round-worms almost exclusively.

Dose, gr. ss–v for adult, and gr. ¼–ss for child.

Officinal Preparations.

Trochisci Santonini, ... 1 = gr. ss.
Sodii Santoninas (Pharm., 1880), . . gr. ij–v.

The sodium santoninate is more soluble, and therefore more dangerous than santonin.

Officinal Name, ASPIDIUM. *English Name*, MALE FERN
—FILIX-MAS.

Definition.—The *rhizome* of Dryopteris filix-mas and Dryopteris marginalis.

Natural Order.—Filices.

Its activity is due to an *oleoresin*, which is officinal.

Dose, ʃʒss–j of oleoresin, taken in the morning after fasting twenty-four hours, and repeat in three hours. Used mostly for tapeworm.

Officinal Name, PEPO. *English Name*, PUMPKIN SEED.

Definition.—The *seed* of Cucurbita pepo.

Natural Order.—Cucurbitaceæ.

Dose, of seed (powdered with sugar), ʒj–ij.

Turpentine, in doses of f𝔷ss, is occasionally given with twice its bulk of castor oil, for both tape- and round worms.

Officinal Name, GRANATUM. *English Name*, POME-
GRANATE.

Definition.—The *bark* of the *stem* and *root* of Punica granatum.

Natural Order.—Lythrarieæ. *Habitat.*—India. Cultivated in United States.

Contains the active alkaloids **pelletierine** and **isopelletierine.**

The decoction (𝔷ij–Oij of water boiled to Oj) is given in doses of f𝔷ij, before breakfast; repeated if necessary.

Dose of pelletierine, gr. v–x.

Officinal Name, KAMALA. *English Name*, KAMALA.

Definition.—The *glands* and *hairs* of the capsules of Mallotus philippinensis.

Natural Order.—Euphorbiaceæ. *Habitat.*—East Indies.

Contains the active resinoid *rottlerin*.

Dose, 𝔷j–ij of powder, given in syrup.

FAMILY III.—DIGESTANTS.

Remedies which increase the action of the gastric and intestinal juices in so far as their solvent power is concerned, and thereby render digestion easier and quicker of accomplishment.

23

Officinal Name, PEPSINUM SACCHARATUM. *English Name*, PEPSIN—SACCHARATED PEPSIN.

Definition.—A proteolytic ferment or enzyme from the fresh stomach of the pig. Capable of digesting not less than 3000 times its own weight of freshly coagulated and disintegrated egg-albumen. This is pure pepsin. When we add 90 per cent. of sugar of milk, we obtain saccharated pepsin. This digests 300 times its weight of albumen.

Dose, gr. v–xx.

Officinal Name, PANCREATINUM. *English Name*, PANCREATIN.

Definition.—A mixture of the enzymes found in the *pancreas* of the Mammalia, that of the hog being generally used.

Dose, gr. v–x.

EXTRACT OF MALT.*

Definition.—The *seeds* of barley (Hordeum distichum, *natural order*, Gramineæ) caused to germinate artificially and then dried. The extract should be free from starch.

Dose, f℈j–f℥j.

PAPAIN.*

Definition.—A ferment from the *fruit* of Carica papaya, a South American tree.

---

* Not officinal.

Natural Order.—Papayaceæ.   *Habitat.*—South America.

Dose, gr. j–x.

FAMILY IV.—ABSORBENTS.

Remedies used to absorb acid and deleterious materials, offensive discharges, secretions, etc., both in the alimentary canal and externally. They are animal and vegetable charcoal.

Officinal Name, CARBO ANIMALIS.   *English Name,* ANIMAL CHARCOAL.

Definition.—Charcoal prepared from bones, blood, etc.

Dose, ℥ss.

Officinal Preparation.

Carbo Animalis Purificatus.

Officinal Name, CARBO LIGNI.   *English Name,* CHARCOAL.

Definition.—Charcoal prepared from soft wood and very finely powdered. Used to dress foul wounds and ulcers, and to absorb noxious gases.

FAMILY V.—DISINFECTANTS.

Substances employed for the prevention of noxious miasmata or effluvia. Of course, heat,—both dry and moist,—air, water, ventilation, and proper disposal of infected excreta are of the utmost import-

ance. Besides these, the various salts of iron and lead, forming sulphurets, are considered important. The oxides of iron convert ordinary oxygen into ozone, which is itself a disinfectant.

COPPERAS—IMPURE SULPHATE OF IRON—

Is an important destructive disinfectant but is not strictly a germicide. Its sole use is to alter the course of putrefaction and destruction of the products thereof. May be used in solution, powder, or crystals, according to the mass to be acted upon. Is extremely valuable for use in cesspools, etc.

LIME

Is of use only as a destructive agent; it tends to prevent odor, but is useless in sewers, privies, etc.

CORROSIVE SUBLIMATE.—BICHLORIDE.—BICHLORIDE OF MERCURY.

Very powerful, both antiseptic and germicidal—even in weak solution. Owing to its poisonous nature care must be observed in its use.

CARBOLIC ACID.

An active germicide. From it we obtain creosols, cresylic acid, creolin, lysol, etc., all also germicides.

Creolin is an emulsion of creosol obtained by means of resin soap.

Lysol is said to contain 50 per cent. of creosol. Both creolin and lysol will mix with water, alcohol, and ether.

PERMANGANATE OF POTASSIUM.

A valuable disinfectant and germicide, but of limited power as it yields up its own oxygen and becomes inert. It will destroy most, if not all, alkaloids in a sufficient length of time, and has been used with success in morphine poisoning, given in doses one-third larger than the amount of morphine ingested.

CHLORINE.

The gas is a powerful germicide, but is exceedingly dangerous, and is apt to injure the clothing as well as the wearer thereof.

Officinal Preparation.

Aqua Chlori contains at least 0.4 per cent. of chlorine gas.

Diluted, may be used as a gargle in diphtheria and as a stimulant in the washing of foul ulcers.

Officinal Name, CALX CHLORATA.   *English Name,*
CHLORINATED LIME.

Definition.—Composed of calcium hypochlorite and calcium chloride. Owes its activity to the chlorine (25 per cent.) it yields when exposed to the air. Used as a disinfectant only.

Officinal Name, LIQUOR SODÆ CHLORATÆ. *English Name*, LABARRAQUE'S SOLUTION.

Definition.—Solution of chlorinated soda.

Contains hypochlorite of soda, and may be used in the same manner as chlorinated lime, or properly diluted in the same way as aqua chlori.

Officinal Name,	*English Name*,
ACIDUM BORICUM.	BORIC OR BORACIC ACID.
SODII BORAS.	BORAX.

Definition.—Used as a dusting powder and in solution for wounds, ulcers, abscesses, burns, etc. Is also efficient in neutralizing ammoniacal urine and in cystitis due thereto, and as an eyewash.

Dose, acid, gr. v–x ; salt, gr. xx.

Officinal Name, ACIDUM SULPHUROSUM. *English Name*, SULPHUROUS ACID.

Definition.—This acid and its salts are very efficient in destroying the low forms of life connected with fermentation and putrefaction, and for this reason form an excellent preservative of organic matter.

Officinal Name, NAPHTALINUM. *English Name*, NAPHTALENE.

Definition.—A hydrocarbon obtained from coal-tar. Poisonous to the lower forms of life. Has supplanted camphor as a destroyer of moths. Also used

externally as an antiseptic dressing and in certain parasitic skin diseases, but beta-naphtol is superior to it, and oftener used.

Dose, gr. ij–viij in capsule.

Officinal Name, NAPHTOL.   *English Name,* BETA-NAPHTOL.

Definition.—Prepared by heating naphtalin with sulphuric acid, then fusing with alkaline hydrates.

Alpha-naphtol is not officinal, but beta-naphtol is.

The odor faintly suggests carbolic acid ; occurs in pale buff or colorless crystals, freely soluble in alcohol, slightly in water. May be used internally as an antiseptic, and externally for the same indications as naphtalene.

Officinal Name, AQUA HYDROGENII DIOXIDI.   *English Name,* HYDROGEN DIOXIDE OR PEROXIDE.

H_2O_2, water, plus one atom of oxygen, is a powerful germicide ; effervesces when brought into contact with pus, destroying the pus-corpuscles ; is a powerful deodorant and an exceedingly valuable local antiseptic. Generally used in diluted solution. Probably not poisonous, but as a matter of precaution the solution used should not be stronger than the officinal solution,—three per cent., by weight, of the pure dioxide,—equal to about ten volumes of available oxygen.

CHAPTER IV.

POISONING.

Antidotes and Treatment.

. In all cases, except those specially mentioned, the stomach should be at once evacuated either by (1) emetic (mustard, salt-water, etc.), (2) hypodermic injection of apomorphinæ, or (3) the stomach-pump.

It should be borne in mind that **tannic acid** is an antidote for all the alkaloids and such vegetable drugs as depend for their activity on the alkaloids they contain. It is also the antidote for antimony, and can do no harm even in large amount.

ÆTHER. ETHER.

Stop the inhalation; artificial respiration, atropine and strychnine hypodermically, fresh air, alternate hot and cold douche, electricity.

ALCOHOL (Acute).

Evacuate the stomach, hot and cold douche, electricity, strychnine hypodermically.

ALKALIES: LIME. POTASSA. SODA.

Weak acid, administered freely and at once. (Vinegar is, as a rule, easily and quickly obtained.)

359

AMMONIUM. AMMONIA.

Neutralize by acid (dilute acetic acid—vinegar—is nearly always at hand); use oils and demulcents to protect the irritated surfaces; opium if necessary. Edema of the glottis or larynx demands tracheotomy.

ANTIMONII ET POTASSII TARTRAS. TARTAR EMETIC.

Evacuate the stomach; give the chemic antidote, tannic acid; opium if needed, and stimulants if required.

ARGENTI NITRAS. NITRATE OF SILVER.

Common salt (sodium chloride) forms the insoluble chloride of silver, which is inert. Treat symptoms as they arise.

ATROPINE. BELLADONNA.

Evacuate stomach (emetics), bowels (cathartics), and bladder (catheter) to prevent further absorption. Tannic acid should be given in large doses. Use morphine hypodermically as a physiologic antidote.

ACIDUM ARSENOSUM. ARSENIC OR ANY OF ITS PREPARATIONS.

Evacuate the stomach. Give the *fresh* antidote (hydrated oxide of iron with magnesia) in large doses.

24

ACIDUM CARBOLICUM. CARBOLIC ACID.

Very rapid in its action. Evacuation of the stomach is useless; use the antidote, a **soluble sulphate,**—preferably magnesia sulphate,—in large amounts, during all stages, as it is capable of neutralizing the poison even after absorption, provided the length of time it has been absorbed has not been too great.

ACIDS.

HYDROCHLORIC OR MURIATIC. NITRIC. SULPHURIC.

First, give alkalies or alkaline earths; soap or *very weak* ammonia solution, to neutralize the acid; then give demulcents, oils, albumens, etc., to protect the injured surfaces and treat the symptoms as they arise.

Hydrochloric acid causes a slight blistering and yellow tinge to surfaces with which it is brought into contact.

Nitric acid causes a persistent orange-yellow stain.

Sulphuric acid blackens and chars the tissues with which it is brought into contact.

ACIDUM HYDROCYANICUM.

An overdose kills almost immediately; treatment is generally useless, but if possible evacuate the stomach; atropine and strychnine hypodermically. Ammonia by the mouth, inhalation, and intraven-

ously. Hot and cold douche, electricity, and artificial respiration.

ACIDUM OXALICUM.

Lime, chalk (calcium carbonate), present in every house as tooth powder, forms an insoluble oxalate and is the *only* antidote. Must be given immediately. If necessary, use the lime from wall or fence.

ACONITUM. ACONITE.

Evacuate the stomach. Keep the head at the same level as the rest of the body. Give alcohol, ether, ammonia, digitalis, hypodermically, if necessary.

CHLORAL. CHLORAL HYDRATE.

Keep up the temperature. Stimulate the heart, use electricity, artificial respiration, and use atropine as required. Chloral is a good antidote in strychnia poisoning, but strychnia is of no use in chloral poisoning.

CHLOROFORM.

Stop the inhalation; artificial respiration, electric battery, atropine, strychnine, etc.

COLCHICUM.

Give emetics and tannic acid. Use opium and stimulants.

COPPER ACETATE.   CUPRUM ACETAS..
COPPER SULPHATE.   CUPRUM SULPHAS.
COPPER ACETATE (IMPURE).   VERDIGRIS.

Potassium ferrocyanide forms an insoluble compound with the above.   Use demulcents, opium to allay pain, and treat the symptoms as they arise.

In chronic poisoning, use iodide of potassium, gr. x, t. i. d. for continued periods of time.

CYANIDE OF POTASSIUM.
CYANIDE OF SILVER.

Treat poisoning the same as hydrocyanic acid.

DIGITALIS.   FOXGLOVE.

Use emetics and cathartics.   Tannic acid is unreliable but is sometimes used.   Give the physiological antidotes alcohol, opium, and ammonia.

HYDRARGYRUM.   MERCURY AND ITS PREPARATIONS.

Albumen (white of egg, milk, or wheat flour) followed by an emetic ; demulcents ; opium if required.

IODUM.   IODINE.

Evacuate the stomach and give liquids containing *starch* in large amount.

Evacuate the stomach ; give tan
antidote.   Use chloral and pota
large doses pushed to effect.   Pr
sis and diuresis ; use the catheter i

OPIUM.   (See Morph

PHOSPHORUS.

French oil of turpentine, whicl
fully used abroad, is practicall
America.   Sulphate of copper fo
sulphate, and therefore is the a
quently employed in this country.
Give gr. x doses and repeat.

PLUMBUM.   LEA

Evacuate the stomach.   Give
or dilute sulphuric acid (forming
phate), demulcents if necessary,
for pain and vomiting.

In chronic lead poisoning, dilute sulphuric acid as a habitual drink, magnesium sulphate as both purge and antidote, and potassium iodide in gr. x doses continued for some weeks will be found of valuable service. Alum is sometimes, though rarely, used.

VERATRUM VIRIDE.

An overdose generally produces emesis and so renders poisoning rare. Treat symptomatically. Give heart stimulants as required.

ZINCI SULPHAS.

Alkalies and alkaline carbonates produce insoluble precipitates.

Give demulcents and treat symptomatically.

ADDENDA.

In order to facilitate the study of the natural orders, the following table is here included:

Algæ, ...	Chondrus.
Anacardieæ, ...	Rhus glabra.
Apocynaceæ, ...	{ Aspidosperma. / Strophanthus.
Aristolochiaceæ, ...	Serpentaria.
Aroideæ, ...	Calamus.
Berberideæ, ...	Podophyllum.
Burseraceæ, ...	Myrrha.
Celastrineæ, ...	Euonymus.
Chenopodiaceæ, ...	Chenopodium.
Coleoptera, ...	Cantharis.

Compositæ, ... {
- Anacyclus pyrethrum.
- Arnicæ.
- Artemisa pauciflora (santonin).
- Anthemis.
- Erigeron.
- Eupatorium.
- Grindelia.
- Lactucarium.
- Tanacetum.
- Taraxacum.

373

Coniferæ, ...	Abies canadensis. Abies excelsa. Juniperus communis. Juniperus sabina. Pinus palustris.
Convolvulaceæ, ...	Jalapa. Scammonium.
Cruciferæ, ...	Sinapis alba and sinapis nigra.
Cucurbitaceæ, ...	Colocynthis. Elaterium. Pepo.
Cupuliferæ, ...	Quercus alba. Quercus lusitanica. Fagus sylvatica (beech).
Ericaceæ, ...	Chimaphila. Uva Ursi.
Euphorbiaceæ, ...	Cascarilla. Kamala. Oleum ricini. Oleum tiglii.
Filices, ...	Aspidium.
Fungi, ...	Ergota. Ustilago.
Gentianeæ, ...	Gentiana. Chirata.
Geraniaceæ, ...	Geranium.
Gramineæ, ...	Rye (secale). Zea mays.
Guttiferæ, ...	Cambogia.
Juglandaceæ, ...	Juglans.
Labiatæ, ...	Hedeoma. Melissa. Mentha piperita. Mentha viridis. Salvia.

Laurineæ, ... $\left\{\begin{array}{l} \text{Camphora.} \\ \text{Cinnamomum zeylanicum.} \\ \text{Sassafras variifolii.} \end{array}\right.$

Leguminosæ, ... $\left\{\begin{array}{l} \text{Acacia.} \\ \text{Balsamum Peruvianum.} \\ \text{Balsamum tolutanum.} \\ \text{Cassia fistula.} \\ \text{Catechu.} \\ \text{Copaiba Langsdorffii.} \\ \text{Glycyrrhiza.} \\ \text{Hæmatoxylon.} \\ \text{Kino.} \\ \text{Physostigma.} \\ \text{Scoparius (sparteine).} \\ \text{Senna.} \\ \text{Tamarindus.} \\ \text{Tragacantha.} \end{array}\right.$

Lichenes, ... Cetraria.

Liliaceæ, ... $\left\{\begin{array}{l} \text{Aloes.} \\ \text{Allium.} \\ \text{Colchicum.} \\ \text{Convallaria majalis.} \\ \text{Sarsaparilla.} \\ \text{Scilla.} \\ \text{Veratrina.} \\ \text{Veratrum viride.} \end{array}\right.$

Lineæ, ... $\left\{\begin{array}{l} \text{Coca.} \\ \text{Linum.} \end{array}\right.$

Lobeliaceæ, ... Lobelia.

Loganiaceæ, ... $\left\{\begin{array}{l} \text{Gelsemium.} \\ \text{Ignatia.} \\ \text{Nux vomica.} \\ \text{Spigelia.} \end{array}\right.$

Lythrarieæ, ... Punica granatum.

Malvaceæ, ... { Althæa.
{ Gossypii.

Menispermaceæ, . . . { Calumba.
{ Pareira.

Myristicaceæ, ... Myristica.

Myrtaceæ, ... { Caryophyllus.
{ Eucalyptus.
{ Oleum cajuputi.
{ Pimenta.

Oleaceæ, ... Manna.

Papaveraceæ, ... { Opium.
{ Sanguinaria.

Papayaceæ, ... Papain.

Piperaceæ, ... { Piper.
{ Matico.
{ Cubeba.

Polygaleæ, ... { Krameria.
{ Senega.

Polygonaceæ, ... Rheum.

Ranunculaceæ, ... { Aconitum.
{ Adonidine.
{ Cimicifuga.
{ Hydrastis.

Rhamnaceæ, ... { Rhamnus frangula.
{ Rhamnus purshiana.

Rosaceæ, ... { Brayera (cusso).
{ Prunus virginiana.
{ Amygdala.
{ Rosa.

Rubiaceæ, ... { Caffea arabica.
{ Cinchona.
{ Ipecacuanha.

Ruminantia, ... Musk.

Rutaceæ, ... {
Aurantii.
Buchu.
Pilocarpus.
Ruta.
}

Scitamineæ, ... {
Cardamomum.
Zingiber.
}

Santalaceæ, ... Oleum santali.

Sapindaceæ, ... Paullinia sorbilis (guarana).

Sterculiaceæ, ... Theobroma.

Styraceæ, ... Benzoinum.

Solanaceæ, ... {
Belladonna.
Capsicum.
Hyoscyamus.
Stramonium.
Tabacum.
}

Thymelæaceæ, ... Mezereum.

Ternstrœmiaceæ, ... Camellia thea (Thea sinensis .

Urticaceæ, ... {
Cannabis indica.
Humulus.
Ulmus.
}

Umbelliferæ, ... {
Ammoniacum.
Anisum.
Asafœtida.
Carum.
Conium.
Coriandrum.
Fœniculum.
Petroselinum sativum.
}

Valerianeæ, ... Valeriana.

Zygophylleæ, ... Guaiaci.

INDEX.

383

Lightning Source UK Ltd.
Milton Keynes UK
UKHW051155041218
333087UK00008BD/859/P